Bruce W. Char   Keith O. Geddes   Gaston H. Gonnet
Benton L. Leong   Michael B. Monagan   Stephen M. Watt

First Leaves:
A Tutorial Introduction
to
Maple V

Springer-Verlag
New York Berlin Heidelberg London Paris
Tokyo Hong Kong Barcelona Budapest

Bruce W. Char
Department of Mathematics
  and Computer Science
Drexel University
Philadelphia, PA 19104
USA

Keith O. Geddes
Department of Computer Science
University of Waterloo
Waterloo, ON
Canada N2L 3G1

Gaston H. Gonnet
Department Informatik
ETH—Zentrum
8092 Zürich
Switzerland

Benton L. Leong
Symbolic Computation Group
University of Waterloo
Waterloo, ON
Canada N2L 3Gl

Michael B. Monagan
Department Informatik
ETH—Zentrum
8092 Zürich
Switzerland

Stephen M. Watt
IBM T. J. Watson Research Center
P.O. Box 218
Route 134
Yorktown Heights, NY 10598
USA

Library of Congress Cataloging-in-Publication Data
First Leaves : a tutorial introduction to Maple V / Bruce W. Char. . .
    [et al.]
          p.   cm.
      Includes bibliographical references and index.
      1. Maple (Computer Program) 2. Algebra—Data processing.
    I. Char, Bruce W.
    QA155.7E4556 1992   91-47590
    510$'$.285$'$369—dc20

Maple is a registered trademark of Waterloo Maple Software.

Printed on acid-free paper.

Photocomposed copy prepared from the authors' TEX files.
Printed and bound by R.R. Donnelley & Sons, Harrisonburg, VA.
Printed in the United States of America.

9 8 7 6

ISBN 0-387-97621-3 Springer-Verlag New York Berlin Heidelberg (hardcover)
ISBN 3-540-97621-3 Springer-Verlag Berlin Heidelberg New York (hardcover)

About Maple: an Overview

0.1 Maple is a system for mathematical computation

Maple is a system for mathematical computation — symbolic, numeric, and graphical. Maple's hundreds of built-in mathematical commands are the product of a decade-long, on-going research project centered at the University of Waterloo and ETH Zurich. Maple (version 3.3) first appeared commercially in 1985. Today, Maple V is in daily use world-wide by scientists, engineers, mathematicians, educators, and students.

First Leaves is a tutorial introduction to Maple and its uses. With the aid of numerous examples, *First Leaves* defines, explains, and illustrates how you can use Maple as a tool for problem-solving and insight.

Let's take a quick look at Maple in action.

Example 1
Potpourri of Maple commands

Compute the sum of cubes of all the integers between 1 and 1234567890987.

```
> S := sum(i^3, i=1..1234567890987);
     S := 580764308853803529746462848150892628716236806084
```

Compute the factors of the sum.

```
> ifactor(S);
        2    2     2      2      2            2           2
     (2)  (3)  (31)  (103)  (449)  (128882753)  (687398603)
```

Solve a system of linear equations in three unknowns and one parameter.

```
> solve({a*x+3*y+3*z=10, x-y+a*z=2, 3*x-2*y+z=6},{x,y,z});
                                                   2
                a - 5            19 a - 5       3 a  - 16 a + 5
     {z = - ---------, x = ---------, y = - ----------------}
            a (4 + a)      a (4 + a)          a (4 + a)
```

Enter a differential equation and label it "de".

```
> de := diff(y(x),x,x) + 2*diff(y(x),x) + y(x) = exp(-x);
            / 2      \
            |  d     |       / d      \
     de :=  |----- y(x)| + 2 |---- y(x)| + y(x) = exp(- x)
            |   2     |       \ dx     /
            \ dx     /
```

Solve the differential equation with initial conditions.

```
> soln := dsolve({de, y(0)=1, D(y)(0)=0}, y(x));
                            2
     soln := y(x) = 1/2 x  exp(- x) + exp(- x) + exp(- x) x
```

Plot the solution (see Figure 1).

```
> plot(rhs(soln),x=0..8);
```

Solve the non-linear equation $\cot(x + \frac{\pi}{8}) = x$ to 40 digits.

```
> Digits := 40;
                              Digits := 40
> fsolve(cot(x+ Pi/8)=x, x);
              .6218013732036774103273149629860812356122
> Digits := 10:
```

Solve a recurrence relation with initial conditions.

```
> reqn := 6*f(n+3) - 103*f(n+2) + 144*f(n+1) = 47*f(n);
    reqn := 6 f(n + 3) - 103 f(n + 2) + 144 f(n + 1) = 47 f(n)
> rsolve({reqn,f(0)=1,f(1)=2,f(2)=4},f(n));
              123   164      n     27        n
              --- - --- (1/2)  + ---- (47/3)
              44    91              4004
```

Compute an indefinite integral:
$\int \frac{8\,x^3+7\,x^2-3}{6\,x^4+5\,x^3-60\,x^2-20\,x-21}\,dx$.

```
> integrand :=(8*x^3+7*x^2-3)/(6*x^4+5*x^3-60*x^2-20*x-21);
                                3       2
                             8 x  + 7 x  - 3
            integrand := -------------------------------
                           4      3       2
                         6 x  + 5 x  - 60 x  - 20 x - 21
> int(integrand,x);
    276              1041                 815        2
    --- ln(x - 3) + ---- ln(2 x + 7) + ----- ln(3 x  + x + 1)
    403              1781                25482
            4387    1/2                            1/2
        + ------ 11      arctan(1/11 (6 x + 1) 11    )
          140151
```

Compute an improper definite integral: $\int_0^\infty e^{-x^2}\ln(x)dx$.

```
> int(exp(-x^2)*ln(x), x=0..infinity);
                  1/2                1/2
          - 1/4 Pi    gamma - 1/2 Pi    ln(2)
```

Define a matrix with numeric entries.

```
> with(linalg):
Warning: new definition for    norm
Warning: new definition for    trace
> A := matrix(4,4,[ -26, -28, 89, -456, 104, 258,
> -770, 3132, 40, 88, -266, 1109, 2, 1, -4, 26]);
                          [ -26  -28    89   -456 ]
                          [                       ]
                          [ 104  258  -770   3132 ]
                   A :=   [                       ]
                          [  40   88  -266   1109 ]
                          [                       ]
                          [   2    1    -4     26 ]
```

Compute its matrix exponential.

```
> evalf(matrixexp(A));
     [  36.63075000    100.0804420   -294.4444646    1171.710326 ]
     [                                                            ]
     [   7.668999383    20.02962192   -59.63774815    240.4005748 ]
     [                                                            ]
     [ -14.84176939    -40.80358790   119.8619492    -476.1546382 ]
     [                                                            ]
     [  -5.503634852   -15.02221644    44.20952586   -175.9809800 ]
```

Define a 3×3 Toeplitz matrix with symbolic entries.

```
> T := toeplitz([a,b,c]);
                          [ a  b  c ]
                          [         ]
                   T := [ b  a  b ]
                          [         ]
                          [ c  b  a ]
```

Compute its eigenvalues.

```
> eigenvals(T);
                                 2        2 1/2
         a - c, a + 1/2 c + 1/2 (c  + 8 b )     ,
                               2        2 1/2
              a + 1/2 c - 1/2 (c  + 8 b )
```

Three dimensional surface plot of $x^2 + 3\,J_0(y^2)e^{1-x^2-y^2}$ (see Figure 2).

```
> plot3d(x^2+3*BesselJ(0,y^2)*exp(1-x^2-y^2),
> x=-2..2, y=-2..2, axes=FRAME);
```

Draw two tubes using three-dimensional surface plotting (see Figure 3).

```
> with(plots):
> tubeplot({[10*cos(t),10*sin(t),0,t=0..2*Pi,
>   radius=2+cos(7*t), numpoints=120, tubepoints=24],
>   [0, 10+ 5*cos(t), 5*sin(t), t=0..2*Pi,
>   radius=1.5, numpoints=50, tubepoints=18]},
>   scaling=CONSTRAINED);
```

More examples of Maple in action can be found in Chapter 6. The basic Maple program, referred to as the *kernel*, can do many simple mathematical calculations involving numbers and formulae. When more complicated calculations are called for, it can draw on a library of mathematical procedures. The Maple library contains directions, specified in Maple's programming language, for doing more complicated things such as factoring polynomials, performing integration, or solving systems of equations. These library procedures are ordinarily kept in files on the disk storage of your computer, but most are automatically loaded into Maple when needed. Together, the kernel and library contain Maple's built-in knowledge of mathematical computations.

There are other mathematical operations for which no "pre-packaged" programming exists. Step-by-step directions must be given to the computer to get the desired result in that case. For example, there is no built-in command to rearrange an equation so that the occurrences of the symbol x and those of y are

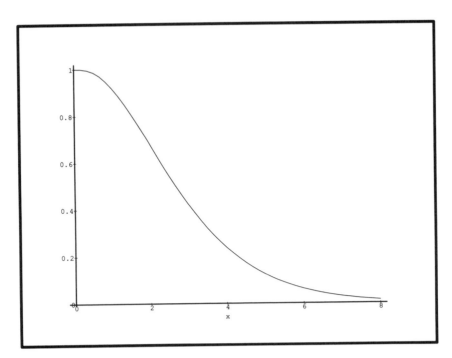

FIGURE 1. A Maple graph of the solution to a differential equation (see Example 1)

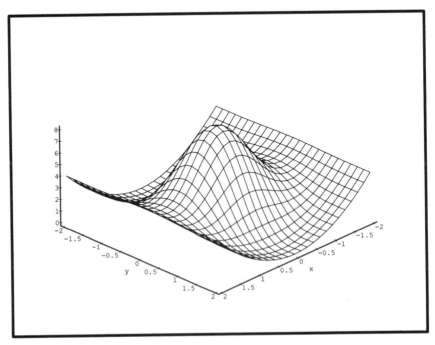

FIGURE 2. A Maple graph of $x^2 + 3\,J_0(y^2)e^{1-x^2-y^2}$

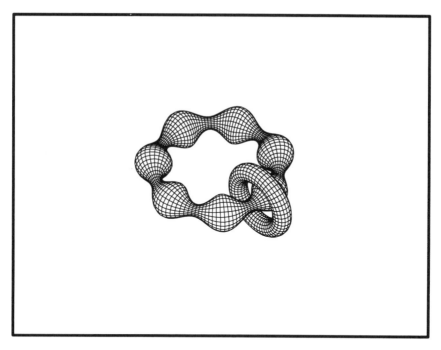

FIGURE 3. A Maple graph of two tubes (see Example 1)

on different sides of the "=" sign. *First Leaves* will explain how to solve problems using Maple that require multiple steps.

For yet larger, more complicated problems, the best approach may be to customize Maple's computing power by writing your own programs in Maple. Even though many useful calculations can be performed without knowledge of programming, you may wish to program Maple so that it knows about mathematical functions not supplied by the library, or to automate a laborious sequence of calculations.

Example 2 illustrates Maple's programming language which is similar in many ways to conventional programming languages such as Pascal. By learning this language, you will be able to program in the same language and have access to the same facilities as those who created the thousands of procedures in Maple's built-in library. After you have learned the language, you may be interested in browsing through some of the programming of Maple library procedures — this is explained in Section 3.17.

Example 2
Fibonacci procedure

Here we define a short Maple program to compute the nth Fibonacci number, defined to be the sum of the previous two Fibonacci numbers. Details of the Maple programming language, such as how to write procedures like this one, are discussed in Chapter 3.

```
> F := proc(n)
>     option remember;
>     if nargs<>1 or not type(n,integer) or n<0 then
>         ERROR(`wrong number or type of parameters`)
>     else
>         if n<2 then n else F(n-1) + F(n-2)   fi
>     fi
> end:
```

Now compute the 10th Fibonacci number by executing the program with $n = 10$.

```
> F(10);
```
$$55$$

Compute a few values in the sequence.

```
> for i from 0 to 2 do
>     lprint(`F(`, 10^i, `) is `, F(10^i))
> od;
F(    1   ) is    1
F(    10  ) is    55
F(    100 ) is    35422484817926191 5075
```

How to use this book: a chapter outline

- Chapter 1: Maple the first time — starting up, doing basic calculations with Maple, and quitting. This chapter describes the most important features of a Maple session.

- Chapter 2: How to derive results that take several steps. This chapter also discusses the internal structure of Maple expressions, which will allow you to access and manipulate pieces and conglomerations of results: leading coefficients, constant terms, the right hand sides of equations, and solution sets.

- Chapter 3: How to develop and use Maple programs. Maple's features for debugging are discussed. You will also learn how to extend the Maple library with packages of your programs.

- Chapter 4: Plotting of functions in two and three dimensions.

- Chapter 5: Efficient computing in Maple. In this chapter, we discuss how to monitor the time and memory Maple uses, how to find the most expensive parts of your programs, and how to speed up numerical computing in Maple.

- Chapter 6: Examples of problem solving with Maple.

- Chapter 7: Access to information – Maple publications, discussion groups, electronic archives, and electronic mail.

- Appendices: Short lists of articles and books on Maple and symbolic computation.

An hour or so spent with Chapter 1 will cover " basic Maple". This should allow you to use many of the most useful symbolic, numerical, and graphing features of Maple. We encourage you to work through the tutorial with both this book and a computer running Maple at hand so that you can experiment with Maple as you read of new commands or concepts.

A more leisurely study of Chapters 2–4 will extend your mastery so that you can solve many common scientific, mathematical, and engineering problems with ease and finesse. The material in Chapter 3 (Maple programming) is especially useful when you have more complicated problems, or special situations not covered by Maple's built-in commands. The material in Chapter 5 can be postponed until you encounter particularly tough computational challenges that require tuning for performance and efficiency. Chapter 6 can be valuable to people of all levels of Maple expertise. It shows how Maple can be applied successfully to problems from the sciences, engineering, and mathematics. It also provides an introduction to Maple problem-solving idioms and tricks that the experts use that often lead to elegant solutions.

Chapter 7 describes how to become part of the international community of Maple users. It also describes the community resources you can retrieve and use electronically. Chapter 7 can be read whenever you wish to pursue such contacts.

In this book, we use the following typographical rules:

- Characters typed at the computer keyboard, and the output Maple responds with are printed in a typewriter font like this:

```
solve(2*x + 1 = 0, x);
```

- Mathematical discussions will use mathematical notation in the standard way: "Let p be the expression $x^2 + 3x + \pi$", for example.

- We will often use *placeholder names* when describing Maple's features. These names are spelled out in an italicized font. For example, `solve(`*equation*`, `*variable*`)`; would be used to indicate the format of Maple's equation solver. It suggests that when using `solve`, you would enter a particular equation and variable to solve, instead of the placeholders *equation* and *variable*.

First Leaves is a complementary work to the *Maple V Language Reference Manual* and the *Maple V Library Reference Manual*, both by Char, Geddes, Gonnet, Leong, Monagan and Watt [CGG+91a; CGG+91b]. The two *Manuals* provide a complete description of Maple in over 900 pages. Obviously, we don't have enough space in *First Leaves* to explain everything! However, *First Leaves* will tell you what you need to know to be a successful problem-solver with Maple. Furthermore, you will "learn how to learn more", so that you can easily extend your mastery of Maple through Maple's ? (help) command or by consulting the *Manuals*.

Acknowledgments

We wish to express our appreciation to the following people who have assisted us with past and present editions of *First Leaves*: Ian Allen, Kate Atherley, Ricardo Baeza-Yates, Robin Carr, William Bauldry, Brooks/Cole Publishing, Stan Devitt, Greg Fee, Eugene Johnson, Bonnie Kent, Leonard Gamberg, Blair Madore, Bev Marshman, Mark Mutrie, Arthur Ogawa, Liyuan Qiao, Tony Scott, John Sellens, Iris Strickler, Waterloo Maple Software, Paul Zorn, and the members of the Symbolic Computation Group of the University of Waterloo. We also appreciate the thoughtful suggestions submitted to us from the readers of the previous versions of *First Leaves* and pre-publication reviewers of this version.

The Maple project has been supported in part by grants from the Academic Development Fund of the University of Waterloo, Digital Equipment Corporation, the Natural Sciences and Engineering Research Council of Canada, the Sloan Foundation, and the Information Technology Research Centre of Ontario.

The authors of *First Leaves* gratefully wish to acknowledge the efforts of all who have participated in the development of Maple. We regret that after a decade of collective effort, we find that we can no longer detail the contributions, accomplishments, and current positions of the many dedicated and talented people involved. The world-wide recognition of Maple's high quality was earned and is deservedly shared by them all.

Contents

The Maple Programming Language   *115*

Advanced Graphics   *155*

Tables

Chapter 1

Interactive Use of Maple

1.1   The user interface and the computational engine

Maple runs on many types of computer systems. Macintosh, DOS (with or without MS Windows), Unix (with or without the X Window System), SunView, NeXT, IBM CMS, and DEC VMS are some that Maple runs on. Nevertheless, all versions of Maple use essentially the same internal programming to perform the mathematical calculations. This part of Maple is known as the *computational engine* or *algebra engine*.

On the other hand, these systems have distinctly different *user interfaces* or *front ends*. The user interface is the part of Maple that handles the entry of information and displays results. Some user interfaces have separate windows for the display of Maple work, and can use a mouse or other pointing device. On other user interfaces, you may have to specify all actions by typing.

In this book, we describe the ways in which you can operate the algebra engine of Maple by typed commands, since most mathematical problem-solving activities involve the use of the algebra engine. Typed commands typically work the same way on all Maple user interfaces. Despite our emphasis on the effects of typed commands, we will indicate where some user interfaces provide alternatives that you may find more convenient.

1.2   Getting started

If you have purchased your own copy of Maple, find the installation instructions and "Getting started" documentation that came with it. Follow the installation instructions if you haven't yet done so.

If you are a first-time user of Maple on institutional or shared equipment, obtain a copy of the site-specific Maple documentation. Such documentation will be available typically from your instructor, or a local computing consultant, administrator, or instructor. This documentation explains how to start Maple on your computer and how the Maple user interface works on it.

In the rest of this book, we refer to this documentation as the "Getting started" documentation, or "the documentation specifically for your computer system". Keep this documentation handy with *First Leaves* as you begin to use Maple.

1.3   Starting a Maple session: how Maple behaves interactively

How you start a Maple session depends on the kind of computer you have, and how Maple was installed on your computer. Table 1 summarizes what typically works. If you have trouble getting Maple started, or if it is acting strangely once you do, re-read your "Getting started" documentation to double-check the details for your computer. If trouble persists, make use of local computing expertise. Also, see Section 7.1 on page 225.

TABLE 1
How to start Maple V on various systems

Type of computer	Typical directions for starting a Maple session*
X Window Systems and DECwindows	Type xmaple.
SunView (Sun user interface)	Type svmaple.
Unix systems	Type maple.
DOS systems	Type maple after seeing the DOS prompt.
Macintosh, NeXT, MS Windows	Select and open ("double-click" on) the Maple icon.

*Be sure to read the Maple documentation specific to your computer system to see if these directions apply to you.

After tending to start-up details that can be handled automatically, Maple prints a *prompt* (typically a > or a •), then positions the *cursor* immediately after it. (For Maple user interfaces that use a pointing device such as a mouse, you can reposition the cursor by moving the pointer and clicking where you wish the Maple cursor to move to.) Once the prompt appears, you can *command* Maple to compute something. As you work with Maple, you will see what you type, and "the answers" — the numbers and formulae that Maple generates in response to your commands.

Whenever you see a prompt, Maple is waiting for you to enter information. Once the computer has started working on an answer, it will not print a prompt until it has finished. On some computers you may see a watch or clock icon while the computer is busy working on a Maple computation.

1.4   Simple arithmetic in Maple

To add, subtract, multiply, or divide integers, fractions, polynomials, or rational functions, type in the expression you wish computed. When working interactively, you then end the expression with a semicolon (;), followed by the return key. (On Macintosh and NeXT systems, use the enter key instead of the return key.)

Example 3
What you see after you type 1 + 1; and press the return key

> 1 + 1;

2

- *ATTENTION*

> **Don't forget the semicolon!** Many users forget this grammatical requirement of Maple when doing their first Maple calculation. You can't progress to the second calculation without the semi-colon ending the first. If you do forget the semicolon, don't panic. Maple will respond with the prompt, indicating that it is still accepting more input, continuing the previous line. You can then enter a semicolon and press return again to complete things.

Table 2 lists Maple's symbols for arithmetic.

TABLE 2
Arithmetic operations in Maple

Operation	Symbol	Example
addition	+	2 + 2
subtraction	−	10 − x
multiplication	*	3 * y * z
division	/	x / 2
truncating division	iquo(...)	iquo(17,3)
integer remainder	irem(...)	irem(10,7)
exponentiation	^ or **	x ^ 2
		x ** 2
absolute value	abs(...)	abs(-4)
factorial	!	10!

- *ATTENTION*

> New Maple users sometimes type 2 x, x2, or x 2 when they mean "multiply 2 and x together". Spaces between symbols are optional, but Maple requires * for multiplication: 2*x or x*2.

After you enter an expression, Maple calculates the result — for example, figuring out that $2 - 3$ is -1. This calculation process is referred to as *evaluating the expression*. Two-dimensional mathematical notation is used to display the result, referred to as *prettyprinted* results. Prettyprinted expressions are found in abundance within the examples of this book. When directing Maple through a sequence of commands, you can refer to the previously computed result by the double-quote symbol (") (see Example 4).[1]   If you don't wish to print a result, you can end things with a colon (:) instead of a semicolon. Example 4 shows some simple arithmetic expressions computed in a Maple session.

You can stretch an expression across several lines; Maple won't compute anything until you finally end the expression with a semicolon and press the return key. On some computers, when you stretch an

---

[1] The symbol ""(two double-quotes) stands for the next-to-last result, and the result before that is referred to by """".

expression over several lines, Maple uses a different kind of prompt such as >> for every line after the first. It reverts to the "primary prompt" when you end entry of the expression.

Example 4
Simple numerical calculations

```
> 2 + abs(-2);
                                        4
```

" refers to the previous result.
```
> (4 + (" * 6))/(999999-32516);
                                       28
                                     ------
                                     967483
```

Here is an input expression that is more than one line long. We use a colon to suppress printing the large number, but we use it in the next calculation with ".
```
> 2 * 3 * 4 * 5 * 6 * 7 * 8 * 9 * 10 * 11 * 12 * 13 * 14 *
>     15 * 16 * 17 * 18:
```

```
> " - 18!;
                                        0
```

1.5   Fixing mistakes

Everyone makes an occasional mistake in typing. If you notice a mistake you've made before you press return, you can just back up and make the correction. If you enter an expression with return and there's a *grammatical* mistake in what you've typed, Maple will respond with a syntax error: *message*. Maple will print out what you typed with an indication of where it noticed an error.

When you see a syntax error message, type in the corrected version of what you want. If you do so, but Maple still complains about syntax errors, then review the Maple documentation describing what you are trying to do. Check closely that you are following the applicable rules. If the documentation includes examples close to what you are trying to do, enter them and see if you can modify them to fit your problem.

Example 5
Syntax errors are often caused by typographical mistakes in entry

A user enters a line missing a few characters. Maple flags this as a syntax error since it doesn't fit its rules for the form of a proper expression. Maple responds with an indication of where it first discovers that something's amiss.

```
> 2 * * 4 * 5 * 6 -;
syntax error:
2 * * 4 * 5 * 6 -;
    ^
```

The user re-enters the line, fixing one mistake but there's still another mistake at the end of the line.

```
> 2 * 3 * 4 * 5 * 6 -;
syntax error:
2 * 3 * 4 * 5 * 6 -;
                  ^
```

A little persistence and success!

```
> 2 * 3 * 4 * 5 * 6 - 240;
                              480
```

Maple will tell you about other sorts of errors besides syntax errors. For example, if you give the command 1/0; Maple responds with `Error, division by zero`. We discuss in more detail the subject of errors and error messages in Section 1.14.

1.6 `help` **yourself to more of Maple**

?topic *?topic,subtopic* *?topic*[*subtopic*] `help(`*topic*`);`	Print out information about *topic*

The ? command allows you to display further details of Maple. (On many workstations and personal computers, this information appears in a separate window.) Using the ? command, you can display descriptions and usage examples of Maple commands and other features. Example 6 shows the result of `?solve`.

?topic is the command for Maple to display a "thumbnail sketch" on the specified topic. As with a dictionary, you must have some idea of where to look for what you are interested in. This requirement is usually fulfilled by knowing the name of the Maple command or topic you wish to find out about, or by consulting the "help index" through `?index`. If you do not hit exactly upon the name of a topic Maple can provide information about, it may suggest one or more alternatives that you can try asking about.

Sometimes it is necessary to display help on a topic by giving two arguments to ? instead of one. For example, as the help displayed in Example 6 suggests, the command ?solve[system] will display information about solving systems of equations. (?solve,system will also work.) ?index,library will list over 250 Maple library functions that do not belong to Maple packages.

Example 6
Information displayed by the ?solve command

```
?solve

FUNCTION: solve - solve equations

CALLING SEQUENCE:
   solve(eqns, vars)

PARAMETERS:
   eqns - an equation or set of equations
   vars - (optional) an unknown or set of unknowns

SYNOPSIS:
- The most common application of solve is to solve a single equation, or to
  solve a system of equations in some unknowns.  A solution to a single equa-
  tion eqns solved for the unknown vars is returned as an expression.  To solve
  a system of equations eqns for unknowns vars, the system is specified as a
  set of equations and a set of unknowns.  The solution is returned as a set of
  equations.

- Multiple solutions are returned as an expression sequence.  Wherever an equa-
  tion is expected, if an expression expr is specified then the equation
  expr = 0 is understood.  If vars is not specified, indets(eqns,name) is used
  in place of vars.

- When solve is unable to find any solutions, the expression NULL is returned.
  This may mean that there are no solutions or that solve was unable to find
  the solutions.

- To assign the solutions to the variables, use the command assign.

- For solving differential equations use dsolve; for purely floating-point
  solutions use fsolve; use isolve to solve for integer roots; msolve to solve
  modulo a prime; rsolve for recurrences, and linalg[linsolve] to solve matrix
  equations.

- Further information is available for the subtopics solve[<subtopic>] where
  <subtopic> is one of

            floats      functions  identity  ineqs    linear
            radical     scalar     series    system
```

- For systems of polynomial equations, the function grobner[gsolve] which uses
  a Grobner-basis approach may be useful.

EXAMPLES:
```
> solve(cos(x) + y = 9, x);
                        Pi - arccos(y - 9)

> solve({cos(x) + y = 9}, {x});
                        {x = Pi - arccos(y - 9)}

> solve(x^3 - 6*x^2 + 11*x - 6, x);
                            1, 2, 3

> solve({x+y=1, 2*x+y=3}, {x,y});
                        {x = 2, y = -1}

> solve({a*x^2*y^2, x-y-1}, {x,y});
                  {x = 1, y = 0}, {x = 0, y = -1}
```

SEE ALSO: dsolve, fsolve, isolve, msolve, rsolve, assign, isolate, match,
            linalg[linsolve], simplex, grobner, solve[<subtopic>] where <subto-
            pic> is one of: floats, functions, identity, ineqs, linear, radical,
            scalar, series, system

? is a special Maple feature: it is a command designed to be used only interactively and cannot be placed inside Maple programs that you will learn to write in Chapter 3. ? does not need a semicolon (;) at the end of the line, unlike all other Maple commands or statements except quit. Maple has another command help(*topic*); that can be used in programs, but it is not as easy to use as ?.

This tutorial, and the *Maple V Language Reference Manual* and the *Maple V Library Reference Manual*, are not revised as often as new versions of Maple appear. You can read a description of the changes to Maple made for the newest version by the command ?updates.

1.7   Parentheses and the priority of arithmetic operations

Entering an expression such as 5 + 4 * 6 into Maple poses a problem in interpretation. Do you mean 5 + 24 (do multiplication first) or 9*6 (do addition first)? Maple applies a set of rules applied uniformly in such situations – its rules for *precedence of operators*. These rules state that multiplication has priority over addition in ambiguous situations such as 5 + 4 * 6, so that Maple would evaluate our example as 29, not 54 . While these rules allow Maple to function in ambiguous situations, you probably will find it less troublesome just to use parentheses in expressions to be clear about what you mean.

- *ATTENTION*

> Parentheses can be used to be clear about the order of arithmetic; e.g., (5 + 4)*6 versus 5 + (4*6). -1^(1/2) means -(1^(1/2)) or -1, so if you wish to refer to the square root of -1, use parentheses: (-1)^(1/2). The precision of meaning conveyed with parentheses usually makes up for the extra typing.

Users who omit "unnecessary" parentheses sometimes encounter problems with the operation of negation. Maple's exponentiation operator requires parentheses as well. $a^{(b^c)}$ and $(a^b)^c$ should be written as a^(b^c) and (a^b)^c, respectively. a^b^c is not a grammatically correct phrase in the Maple language. If you wish to write Maple expressions in the fewest number of parentheses, consult the *Maple V Language Reference Manual*. Using the command ?operators,precedence provides a detailed description of the precedence of operations.

Example 7
Problems that occur when you don't use parentheses liberally

Negation has a lower precedence than exponentiation.	`> -5^2;`
	-25

You must parenthesize negative terms according to the grammatical rules of Maple.	`> -5*-2;`
	`on line 9, syntax error:`
	`-5*-2;`
	`    ^`
	
	`> -5*(-2);`

	10

You must parenthesize towers of exponents.	`> 2^3^4;`
	`on line 14, syntax error:`
	`2^3^4;`
	`   ^`
	
	`> 2^(3^4);`

	2417851639229258349412352

1.8   Ending a Maple session

`quit`
`done`
`stop`
	Terminate a Maple session

TABLE 3
How to conclude a Maple session

Computer System	How to conclude a Maple session
All	Type quit followed by return. (Use enter instead of return on Macintosh or NeXT systems).
Macintosh	Select **Quit** from the **File** menu, or press command-Q.
NeXT	Select the **Quit** menu item, or press command-Q.
X Window System and DECwindows (xmaple) user interface; SunView (svmaple) user interface	Click on the **Quit** button or select **Quit** from the **File** menu.
DOS	Press the F3 key.
MS Windows for DOS	Select the **Exit** menu item or press Alt-F4.

Entering quit, followed by return (enter on a Macintosh or NeXT system), ends a Maple session. done or stop are synonyms for quit. These commands work for all versions of Maple. There are other ways of quitting that work only for specific kinds of computers. Table 3 lists some of these alternatives for Maple V. Consult the system-specific documentation to learn more about quitting your version of Maple.

On some systems, Maple will print out a final message summarizing the computer resources used during the session, as illustrated in Example 8. We return to these "bytes used" messages in Chapter 5.

Example 8
What's printed when departing from Maple on some systems

```
> quit
bytes used=139632, alloc=131048, time=.233
```

1.9   Maple variables

variable := *expression* ;	Assign a value to a variable

You can label a calculated result for further use, as was done with S := sum(i^3, i=1..1234567890987); in Example 1. This is called *assigning a Maple variable a value*. We use the terms *programming variable* and *label for a result* interchangeably. But unlike many conventional programming languages, Maple can use variables in the sense of "a mathematical unknown" as well. A *mathematical variable* or *unassigned variable* is a variable with no assigned value. Mathematical variables represent "algebraic unknowns," as with x and y in Example 1. If you do not wish to sort out which Maple variables have assigned values and which ones don't, then you can consider Maple mathematical variables as having their name as their assigned value.

A Maple variable ordinarily starts with a letter, followed by up to 498 letters, digits, or underscores. Of course, typical variable names aren't close to the maximum length; they are convenient to type and to remember. Weird (but just as valid) Maple names can be created by surrounding characters in backquotes (`` ` ``), also called "left quotes". Names in backquotes can include any character from the keyboard (including blanks), not just letters, digits, and underscores. These are a few examples of valid Maple names:

```
expr1      x                  T                Formula_47
d2DYDX     `a funny name? `   `The answer is= `    `SessionLog.m`
```

- *ATTENTION*

    1. Maple distinguishes between lowercase and uppercase letters. For example, on most computers the name xyz is different from xyZ.
    2. The backquotes mark the boundaries of the name but are not part of it. Thus, Maple considers `` `x` `` to mean the same thing as x. The name `` `1` `` is different from 1, however. The latter is a number, not a name, and it can't be assigned a value or stand for an algebraic unknown. The former can be used either as a programming or mathematical variable. It would make life very confusing for anyone trying to read printed or displayed results, though!
    3. The meaning of the mark `` ` `` is entirely different from that of the mark ', which is an apostrophe, or "right quote" (see Section 2.5). Be careful to use `` ` `` when you mean to!
    4. Actually, it's possible to start a variable name with an underscore character (_) instead of a letter. However, Maple's internal programming uses such "underscore variables". They also sometimes appear as system-generated symbols or constants in solutions to algebraic or differential equations. To avoid disappointment, you should refrain from using such names yourself.

Example 9
Assignment of values (labeling)

force is a label for a famous
formula.

```
> force := mass * acceleration;
                    force := mass acceleration
```

We now assign mass and
acceleration particular values.

```
> mass := 3000;
                    mass := 3000
> acceleration := 9.8;
                    acceleration := 9.8
```

The value of force now
incorporates the assigned values
of mass and acceleration.

```
> force;
                    29400.0
```

If we change the value of `mass` or `acceleration`, the value of `force` reflects the change as well.

```
> mass := 3500;
                              mass := 3500
> force;
                                34300.0
```

Another example, showing one use for a variable which includes blank spaces in its name.

```
> distance1 := 10;
                            distance1 := 10
> distance2 := 50;
                            distance2 := 50
> `Total energy` := force * (distance2 - distance1);
                                              7
                 Total energy := .13720000*10
```

1.10   Built-in commands for mathematical computation

Maple "knows" about certain mathematical objects — numbers, symbolic constants (π, e, etc.), multivariate polynomials and rational functions, and expressions involving standard mathematical functions. An obvious way to see this is to notice that Maple automatically *simplifies* expressions: $\cos(2\pi)$ to 1, or $e^{2\log\sqrt{2}}$ to 2.

Tables 4, 5 and 6 list some of the mathematical constants and functions that Maple knows about. Example 10 shows some of Maple's automatic mathematical simplifications.

TABLE 4
Mathematical constants known to Maple

Constant	Maple name
integers	−47, 1, 2
rational numbers	3/5, −1/3
floating-point numbers	1.0, .002, .35 * 10^(−45), Float(314,−2), Float(−8,5)
true, false	true, false
π	Pi
e (natural log base)	exp(1), E
$\sqrt{-1}$	I, (−1)^(1/2)
∞	infinity
Euler's constant γ	gamma
Catalan's constant $c = \sum_{i=0}^{\infty} \frac{(-1)^i}{(2\,i+1)^2}$	Catalan

TABLE 5
Mathematical functions known to Maple

Function	Maple name
e^x (exponential)	`exp(x)`
$\ln x$ (natural logarithm)	`ln(x)` or `log(x)`
$\log_{10} x$ (logarithm to the base 10)	`log10(x)`
$\log_b x$ (logarithm to the base b)	`log[b](x)`
$\sqrt{x}$ (square root)	`sqrt(x)`
$\|x\|$ (absolute value)	`abs(x)`
Minimum, maximum	`min, max`
Round x to the nearest integer	`round(x)`
Truncate x to its integer part	`trunc(x)`
Fractional part of x	`frac(x)`
Greatest common divisor	`gcd`
Least common multiple	`lcm`
Signum function $$\text{signum}(z) = \begin{cases} 1 & \text{for real } z \geq 0 \\ -1 & \text{for real } z < 0 \\ z/\|z\| & \text{otherwise} \end{cases}$$	`signum(z)`
Sine, cosine, tangent*	`sin(x), cos(x), tan(x)`
Secant, cosecant, cotangent	`sec(x), csc(x), cot(x)`
Inverse trigonometric functions	`arcsin(x), arccos(x), arctan(x),` `arcsec(x), arccsc(x), arccot(x)`
Hyperbolic functions	`sinh(x), cosh(x), tanh(x),` `sech(x), csch(x), coth(x)`
Inverse hyperbolic functions	`arcsinh(x), arccosh(x), arctanh(x)` `arcsech(x), arccsch(x), arccoth(x)`
$n!$, $\Gamma(x)$ (generalized factorial)	`n!, GAMMA(x)`
$\binom{n}{m}$ (binomial coefficient)	`binomial(n,m)`

*Trigonometric functions use radians, not degrees.

● *ATTENTION*

> Maple spells π with a capital P and a lower case i. `PI` and `pi` are not substitutes for `Pi`! Similarly, use `I` and not `i` for the imaginary number, and `exp(x)` for e^x. Maple considers `sqrt(-1)` and `I` to be synonyms.

Example 10
Assortment of automatic mathematical simplifications

$2|\sin(-\frac{\pi}{2})|$ is simplified automatically.

```
> 2*abs(sin(-Pi/2));
```
                                                                                              2

TABLE 6
More mathematical functions known to Maple

Function*	Maple name
Modular arithmetic $a \bmod b$	`a mod b` or `` `mod`(a,b) ``
positive representation†	`modp(a,b)`
symmetric representation†	`mods(a,b)`
gcd of a and b mod p	`Gcd(a,b) mod p`
$a^b \bmod p$	`Power(a,b) mod p` or `a &^ b mod p`
n^{th} Bernoulli numbers	`bernoulli(n)`
n^{th} Euler numbers	`euler(n)`
Gamma function $\Gamma(x) = \int_0^\infty e^{-t}t^{x-1}dt$	`GAMMA(x)`
Beta function $\beta(x,y) = \Gamma(x)\Gamma(y)/\Gamma(x+y)$	`Beta(x,y)`
Derivatives of $\ln\Gamma$ (polygamma functions)	
$\psi(x), \psi'(x), \psi''(x)\ldots$	`Psi(x), Psi(1,x), Psi(2,x) ...`
Error function $\mathrm{erf}(x) = \int_0^x 2e^{-t^2}/\sqrt{\pi}\,dt$	`erf(x)`
Airy wave functions $Ai(x), Bi(x)$	`Ai(x), Bi(x)`
Riemann ζ function (and derivatives)	`Zeta(x), Zeta(1,x) ...`
Dilogarithm function $\int_1^x \ln t/(1-t)dt$	`dilog(x)`
Bessel functions	
$J_v(x), Y_v(x)$	`BesselJ(v,x), BesselY(v,x),`
$I_v(x), K_v(x)$	`BesselI(v,x), BesselK(v,x)`
Exponential/trigonometric integrals	`Ei(x), Ci(x), Si(x)`
Fresnel integrals	`FresnelC(x), FresnelS(x)`
Hypergeometric functions	`hypergeom([],[],z)`
Dirac function $\delta(t)$	`Dirac(t)`
$\delta'(t), \delta''(t)\ldots$	`Dirac(1,x), Dirac(2,x) ...`
Heaviside unit step function $S(t)$	`Heaviside(t)`
$S(t) = 0$ for $t \le 0$, $S(t) = 1$ for $t \ge 0$	

*See the *Maple V Library Reference Manual*, or type `?inifcns` in Maple for a comprehensive list of functions, and for further details. You can also use `?numtheory`, `?combinat`, `?stats`, and `?orthopoly` to see functions that are provided by some of the packages in Maple. You may also wish to consult standard reference works on mathematical functions such as *Handbook of Mathematical Functions*, M. Abramowitz and I. Stegun, eds. [AS65].
†In the positive representation, results range from 0 to $b-1$. In symmetric representation, results are in the range $-\lfloor\frac{|b|-1}{2}\rfloor..\lfloor\frac{|b|}{2}\rfloor$.

$\Gamma(7) - 6!$

```
> GAMMA(7)-6!;
```
$$0$$

$e^{2\ln x}$

```
> exp(2*ln(x));
```
$$x^2$$

In Maple, I and `sqrt(-1)` are synonymous.

```
> I^2*exp(2*Pi*sqrt(-1));
                                           -1
```

Maple's built-in knowledge extends to performing numerical evaluation or approximation, differentiation, integration, solving equations, and finding limits involving mathematical expressions. You can tap this knowledge by giving *commands* to Maple to perform the desired operation: `diff` to differentiate, `solve` to solve equations, etc. In the next sections, we examine Maple's most used commands for mathematical calculation.

You may have noticed that in Example 10 Maple accepted and processed the expression $e^{2\ln x}$, even though x had no assigned value. Maple is designed to handle mathematical operations involving symbols (variables with no assigned values) as well as numbers.

Example 11
Maple calculations with mathematical unknowns

Maple expressions can involve symbolic unknowns as well as numbers. Such expressions can be labeled, just as with numerical results. Maple automatically performs a limited amount of simplification as it evaluates the expression. We explain how to simplify `expr2` further in Section 1.19, which starts on page 38.

```
> expr1 := (x + y + x + x*x*x + x) / 2;
                                                             3
                      expr1 := 3/2 x + 1/2 y + 1/2 x
> expr2 := ((4*expr1) - 2*y) / x;
                                                        3
                                      6 x + 2 x
                      expr2 := ----------
                                          x
```

Symbolic expressions can include any function.

```
> bigexpr := sin( expr1 ) / ln ( expr2 ^ 2 );
                                                                3
                        sin(3/2 x + 1/2 y + 1/2 x )
           bigexpr := --------------------------
                                                 3 2
                                  (6 x + 2 x )
                         ln(------------)
                                        2
                                       x
```

While expressions involving just numbers are always fully reduced to the simplest fraction, Example 11 shows that Maple's automatic procedures for reduction and simplification of symbolic expressions do not always put expressions in the "best" form. In the next section, we learn about symbolic calculations Maple can perform that go beyond just adding or multiplying them together. In Section 1.19, we return to the problem of how to simplify a symbolic answer into the form you wish for it.

1.11 Introducing Maple's mathematical commands

Maple has preprogrammed knowledge of common mathematical operations beyond its knowledge of arithmetic and simplification. Entering *function*(*expression*); — for example, `abs(-55)`; — is the way to tap Maple's computational power. Sometimes the operation involves two or more expressions. Thus, you could compute the greatest common divisor of 55 and 22 with `gcd(55,22)`;. In addition to mathematical functions such as `gcd` or `abs`, Maple has many "computer functions" that have a similar form. Maple functions take one or more expressions as *arguments* to the function and return a single object (number, expression, etc.) as their *value* or *result*. For example, Maple has a function `solve`. The result of evaluating `solve(3*x-5=x, x)`; is 5/2 — what you get when you solve the equation for x.

In this section, we take a brief tour of some commonly used computational functions, including those for solving equations, doing calculus, and plotting graphs. We also will look at `subs`, a useful function for evaluating expressions at a point and for applying identities to expressions. (We discuss Maple's facilities for matrix and vector linear algebra in detail in Chapter 2.)

1.11.1 Solution of algebraic equations

`solve(` *equation* `)` `solve(` *equation*, *variable* `)`	Solve an equation for an unknown

In its simplest forms, `solve` takes an equation in one unknown for an argument. `solve(3*x + 4 = 5*x)` is an example of this. `solve` can be an equation in several unknowns as well, with the request to solve for one of the unknowns in terms of the others. An example of this is `solve(ln(x^2-1) = a, x)`. The value of `solve` will be the number or expression that is the solution of the equation.

If `solve` finds more than one solution, you will see the solutions separated by commas — an *expression sequence*. When the equation lacks solutions, `solve` returns "nothing" (the `NULL` expression sequence, also discussed in Section 2.9). By assigning the result of `solve` to a programming variable, say, `s`, you can refer to a particular component of a multiple solution using a *selection operation* or *indexing* notation (e.g., `s[1]`, `s[2]`, etc.). We discuss expression sequences further in Section 2.9. Further discussion of how to get at components of structured expressions can be found in Sections 2.8 and 2.9.2.

Example 12
Solution of one equation and one unknown

Solve a single equation with one unknown.

```
> solve(3*x + 4 = 5*x);
                                2
```

Solve a single equation with more than one solution. This gives a sequence of solutions.

```
> solve(ln(x^2 - 1) = a, x);
                      1/2                      1/2
         - 1/2 (4 exp(a) + 4)   , 1/2 (4 exp(a) + 4)
```

When an expression is given
instead of an equation, an "= 0"
is implicitly assumed.

```
> solve(q^3-k, q);
   1/3            1/3           1/2  1/3
  k    , - 1/2 k     + 1/2 3    k      I,
              1/3           1/2  1/3
      - 1/2 k     - 1/2 3    k      I
```

One can assign the result of
solve to a programming
variable and then refer to each
solution by an index.

```
> solutions:=";
solutions :=
   1/3            1/3           1/2  1/3
  k    , - 1/2 k     + 1/2 3    k      I,
              1/3           1/2  1/3
      - 1/2 k     - 1/2 3    k      I
> solutions[1];
                                  1/3
                                 k

> solutions[3];
              1/3           1/2  1/3
      - 1/2 k     - 1/2 3    k      I
```

`solve(` *set of equations*, *set of variables* `)`	Solve a set of equations for a set of unknowns

If you give `solve` a set of equations and a set of unknowns to solve for, it will return a set of solutions
(or a sequence of several sets, if there is more than one solution). For example,

```
solve( {a = b + 2, b - a = t}, {a, b} );
```

is a command to solve the system of two equations for the unknowns *a* and *b*.[2]

Each element of the solution set will be an equation of the form *var* = *expr*, where *var* is a variable being
solved for.[3] When a variable being solved for can take on any value in the solution, the solution set includes
a "tautological equation" (e.g., x = x).

Example 13
Solution of a system of equations

We solve a set of linear
equations. The variables to be
solved for are also collected
together into a set. A solution set
is returned.

```
> eqnset := {x+y=b,   a*x-2/3*y = k};
               eqnset := {x + y = b, a x - 2/3 y = k}
> varset := {x,y};
                           varset := {y, x}
> SolutionSet := solve(eqnset, varset);
                        2 b + 3 k        - k + b a
         SolutionSet := {x = ---------, y = 3 ---------}
                          3 a + 2          3 a + 2
```

---

[2]If you have trouble finding { and } on your keyboard, try (* and *) instead.
[3]We discuss sets and equations as data structures in more detail in Section 2.8.

The built-in `assign` function will take the solution set from `solve`, and assign the specified variables accordingly.

```
> assign(SolutionSet);
> x;
```
$$\frac{2\,b + 3\,k}{3\,a + 2}$$

```
> y;
```
$$3\,\frac{-k + b\,a}{3\,a + 2}$$

The values of the equations now include the updated information about x and y. `simplify` is a command which applies its basic knowledge about simplification of expressions. It is discussed further in Section 1.20.

```
> eqnset;
```
$$\{\frac{2\,b + 3\,k}{3\,a + 2} + 3\,\frac{-k + b\,a}{3\,a + 2} = b, \ \frac{a\,(2\,b + 3\,k)}{3\,a + 2} - 2\,\frac{-k + b\,a}{3\,a + 2} = k\}$$

```
> simplify(eqnset);
```
$$\{b = b, \ k = k\}$$

Unassign the names x and y so that their values will again become mathematical unknowns (no assigned value). "Unassignment" is discussed fully in Chapter 3.

```
> x := 'x';   y := 'y';
```
$$x := x$$
$$y := y$$

The equations always reflect this most current information about x and y.

```
> eqnset;
```
$$\{x + y = b, \ a\,x - 2/3\,y = k\}$$

This is the solution of a non-linear system of polynomials with four solutions, generated by picking independently one of two solutions for y and one for x.

```
> solve({y^2+1=x, x+2=y}, {x,y});
```
$$\{y = RootOf(\_Z^2 + 3 - \_Z), \ x = -2 + RootOf(\_Z^2 + 3 - \_Z)\}$$

Maple will always find the solution to a system of linear equations should the solution exist (given enough computer time and memory for computing). However, its methods for nonlinear equations do not always succeed even when a solution exists. If it cannot find any solutions, it will not print out anything at all, just giving a prompt for the next command[4]. Even if Maple finds some solutions to an equation or system of equations, it may not find *all* of them.

Example 14
What happens when there are no solutions, or only some solutions are found

Nothing is printed when no solution can be found.	```\n> solve(x=x+1, x);\n> solve({cos(x*Pi)=0, cot(x*Pi/3)=0}, {x});\n```

Maple does not always give all solutions.

```
> solve(cos(x*Pi)=0,x);
```
$$1/2$$

If no variables are supplied, then Maple will solve for all unknowns.

```
> solve( a*x^2 + b*x + c = 0 );
```
$$\{c = - a\ x^2 - b\ x,\ x = x,\ a = a,\ b = b\}$$

Sometimes a solution involves *algebraic numbers*.[5] Some algebraic numbers have simple forms: 4 and -4 are the zeros of $x^2 - 16$; $\sqrt{2}$ is a zero of $z^2 - 2$. However, sometimes the simplest way to refer to an algebraic number is by referring to it as the root of a specific polynomial equation. Maple refers to the three roots of $x^3 + 3x = 2$ (approximately $.596$ and $-.298 \pm 1.81i$) as

```
RootOf(_Z^3 + 3*_Z - 2).
```

When you see a "RootOf" expression in a result returned by `solve`, it is a placeholder for several solutions—one for each possible root of the polynomial described in the "RootOf."[6] In Example 13, the result of

```
solve({y^2 + 1 = x, x+2=y}, {x,y});
```

is

---

[4]Maple returns NULL, the "null sequence" when it cannot find any solutions. NULL prettyprints as "nothing at all". See Section 2.9 for a further discussion of NULL and sequences.
[5]An algebraic number is a real or complex number that satisfies a polynomial equation with integer coefficients.
[6]Maple's `simplify` command is useful in reducing the size of solutions that have RootOfs in them. See Section 1.19.

$$\{y = \text{RootOf}(\_Z^2 + 3 - \_Z), \ x = -2 + \text{RootOf}(\_Z^2 + 3 - \_Z)\}.$$

This is actually two distinct solutions rolled into one. Since the two roots of $x^2 + 3 - x = 0$ are $(1 \pm \sqrt{11}i)/2,$[7] the two solutions to the system of equations are

$$y = (1 + \sqrt{11}i)/2, \quad x = -2 + (1 + \sqrt{11}i)/2$$

and

$$y = (1 - \sqrt{11}i)/2, \quad x = -2 + (1 - \sqrt{11}i)/2.$$

When you want to generate multiple solutions from a `solve` result with a `RootOf` in it, you must be consistent in the choice of roots if the same `RootOf` occurs several times in the result.

Sometimes it is impossible, or very expensive, to find an exact solution. Maple's `fsolve` procedure, discussed in Section 1.12.2, will use numerical methods to find approximate roots.

1.11.2   Expression substitution with `subs`

`subs( var = replacement, expression )`	Substitute for a variable in an expression
`subs( var`$_1$` = replacement`$_1$`, var`$_2$` = replacement`$_2$`, expression )`	Multiple substitution

In a calculation of several steps, a known identity such as $f = m * a$, $v = i * r$, or $x = 5/3$ often needs to be applied to an expression. We can do this by replacing every f in the expression by $m * a$, or every v by $i * r$, and so on. This can be done in Maple with the first form of the `subs` command. The result is a new expression where every occurrence of the symbol *var* in *expression* is replaced by *replacement*.

A double substitution can be specified by the second form of the `subs` command. Every occurrence of *var*$_1$ is replaced by *replacement*$_1$, then every occurrence of *var*$_2$ is replaced by *replacement*$_2$. `subs` can be used in a similar way to specify multiple (three, four, or more) substitutions: a sequence of variable-replacement equations, followed by the expression to be operated upon.

Symbol replacement is not the only use for `subs`. For example, it can be used to replace one term of a sum with another expression. Knowing precisely what subexpressions can be replaced within an expression involves an understanding of which subexpressions are accessible by the `op` function. Section 2.8 contains a further discussion of `op`.

Example 15
Substitution of an expression for another via `subs`

The subs (substitute) command can be used to take solutions and "plug them into" the original equation to verify the solutions.

```
> soln:= solve(3*x-y=x,x);
                    soln := 1/2 y
> subs(x=soln, 3*x-y=x);
                    1/2 y = 1/2 y
```

---

[7]It doesn't matter whether we use x or $\_Z$ as the variable when describing things to ourselves. However, Maple prefers using the variable named $\_Z$ in its `RootOf`s.

Multiple replacement of symbols by other symbols or expressions.

```
> msol := solve( e=m*c^2,m);
```

$$msol := \frac{e}{c^2}$$

```
> subs(m=msol, a=9.8, c=300000, f = m * a);
```

$$f = .1088888889*10^{-9}\ e$$

Note that the replacements occur in order from left to right as they are specified. In this (misguided) calculation, the substitution of c occurs before there is any c in the expression to replace!

```
> subs(c=300000, m=msol, a=9.8, f= m * a);
```

$$f = 9.8\ \frac{e}{c^2}$$

This shows replacement of an expression by another expression. Replace tangents by a ratio of sines and cosines, and then replace the squared cosines.

```
> expr2 := cos(x)^2 + tan(x)^2;
```

$$expr2 := cos(x)^2 + tan(x)^2$$

```
> subs(tan(x)=sin(x)/cos(x), cos(x)^2=1-sin(x)^2, expr2);
```

$$1 - sin(x)^2 + \frac{sin(x)^2}{cos(x)^2}$$

Well, that almost worked. The quotient with $cos(x)^2$ is being regarded by Maple as being $cos(x)^{-2}$ in a product.

```
> subs(cos(x)^(-2)=1/(1-sin(x)^2),");
```

$$1 - sin(x)^2 + \frac{sin(x)^2}{1 - sin(x)^2}$$

The first argument to subs also can be a set (or a list) of equations. Thus, you can use the result of solve for substitutions even when the result is a set. Giving a set or list of substitutions causes them to be performed simultaneously instead of sequentially.

Example 16
Simultaneous substitution of a set of expressions

```
> f := x*y^2 + z;
```

$$f := x y^2 + z$$

Substitute first for x and then for y.

> subs(x=y, y=x, f);

$$x^3 + z$$

Instead of substituting for the variables in a sequential fashion, substitute for both x and y simultaneously.

> subs({x=y, y=x}, f);

$$y x^2 + z$$

1.12   Using Maple as a numerical calculator

Maple's primary mode of operation uses "exact numbers" – integers and rational numbers of any magnitude. These are different from the *floating-point numbers* used by calculators and conventional computer programming languages such as Fortran, C, or Pascal. Conventional floating-point numbers are expressed with a decimal point and a limited number of digits of precision. They take less memory than Maple exact numbers, and arithmetic with them is typically faster. However, Maple's results with exact numbers are free of the rounding error that is typically unavoidable with fixed-precision floating-point numbers.

There are times when faster-but-less-precise arithmetic is desirable, such as when you want just a rough estimate for an answer and believe that rounding error won't sabotage the quality of the computation. In this section, we present Maple facilities for conventional numerical calculations with floating-point numbers. Maple adds a twist to numerical arithmetic — in Maple you can specify the number of digits used in each step of the calculation. You can ask Maple to be a 10-digit calculator, a 20-digit one, or even a calculator with 300 digits of accuracy. This has a price — Maple is faster at arithmetic with the precision of 10 digits than when doing arithmetic with 100 digits.

1.12.1   Numerical calculations to any desired precision

`evalf( ` *expression* ` )`	Find a numerical approximation
`evalf( ` *expression* `, ` *digits* ` )`	Find a numerical approximation calculating with more digits

Floating-point arithmetic in Maple simulates a calculator with 10 decimal digits of accuracy by default. These numbers are printed as a decimal fraction times a power of 10 — standard "scientific notation". A more natural fixed-point notation is used if the number is within a few orders of magnitude of 1. Maple will convert mixed expressions involving both floating-point and rational numbers into floating-point notation.

Maple differs from ordinary calculators in that you can *increase* or *decrease* the number of digits used in the floating-point arithmetic. The value of the variable `Digits` controls the number of (decimal) digits used during the calculation. It is initially 10, but you can change `Digits` by assigning it a different value, just as you would any other Maple variable. For example, if you wanted a 40-digit calculator, you'd assign `Digits` the value 40 before starting your calculations.

Example 17
Floating-point numbers in Maple

Floating-point arithmetic is used
when an expression contains
floating-point and rational
numbers.

```
> 1.0 + 3/5;
                                  1.600000000
> 3.0*10^20 - 2.99*10^13;
                                                          21
                                  .2999999701*10
```

Floating-point calculations are
not always exact.

```
> 1.0/3.0 + 1.0/3.0 + 1.0/3.0;
                                  .9999999999
> " - 1;
                                                   -9
                                  -.1*10
```

However, approximation can be
a way of conveying useful
information that a similar
rational calculation can obscure.

```
> sum(1/(2*i-1),i=1..10);
                                  31037876
                                  --------
                                  14549535
> sum(1.0/(2*i-1),i=1..10);
                                  2.133255530
```

The value of `Digits` controls the
accuracy of floating-point
calculation.

```
> Digits;
                                             10
> Digits := 5:
> 2.0/3.0 + 5;
                                          5.6667
> Digits := 40:
> 2.0^(1/3);
                   1.259921049894873164767210607278228350570
> Digits := 10:
```

Floating-point numbers are different from rational numbers in Maple because no attempt is made always to keep the answer exact. The amount of work needed for floating-point calculation can be considerably less than for its exact rational counterpart. Thus, floating-point numbers are useful where approximation is good enough to solve the problem.

You may notice that Maple floating-point arithmetic typically runs more slowly than the fixed-accuracy floating-point ("real") arithmetic of languages such as Pascal, Fortran, Lisp, or Prolog. This is due in part to the "adjustable accuracy" feature of Maple's arithmetic.

`evalf` (*expression*) ("**eval**uate in a floating-point context") will invoke procedures from the Maple library to approximate the numeric *expression*. You can include in *expression* any of the Maple operations, functions, and constants shown in Tables 2 through 5 (and others described in the *Maple V Language Reference Manual*).

Example 18
Approximating exact results with `evalf`

`evalf` can be used to force evaluation using floating-point arithmetic whenever functions or expressions containing only rational numbers or exact constants appear.

```
> expr1 := tan(1.0) + tan(2);
                    expr1 := 1.557407725 + tan(2)
> evalf(expr1);
                          -.627632138
> Digits := 20:
> evalf(expr1);
                      -.6276321382615189916
```

You can temporarily override the value of `Digits` for a particular `evalf` computation by including the number of digits as a second argument to `evalf`.

```
> evalf(sqrt(2), 40);
        1.414213562373095048801688724209698078570
```

`evalf` of a `RootOf` gives an approximation to an arbitrarily chosen real root of the polynomial, if there is one.

```
> {solve( (x^3+1) * (x^6+3*x+1), x )};
                       1/2              1/2
   {-1, 1/2 + 1/2 I 3    , 1/2 - 1/2 I 3    ,
                  6
      RootOf(_Z  + 3 _Z + 1)}
> evalf(");
      {-1., .50000000000000000000 + .86602540378443864675 I,
            .50000000000000000000 - .86602540378443864675 I,
             -.33379438854319331628}
```

1.12.2   Numerical solution of equations

`fsolve( expression )`	Solve equations numerically
`fsolve( set of equations, set of variables )`	
`fsolve( set of equations, set of variables, options )`	

Numerical methods for solving mathematical problems typically involve calculations using floating-point arithmetic. `fsolve`, like `solve`, finds solutions to algebraic equations, but it uses numerical methods instead of formula manipulation. While `solve` might use the elimination of variables from equations of a system, the quadratic formula, or polynomial factorization to find solutions, `fsolve` might apply a technique that produces a series of values that converge to the true solution. Since it uses only floating-point arithmetic, the solutions `fsolve` produce can only approximate the exact solutions. By default, `fsolve` will find one real-valued solution to the equation. The exception to this is if *expression* is a polynomial, in which case all real-valued solutions will be returned. `fsolve` will return the input expression when it cannot find a solution, but there still could be a possibility of a solution.

fsolve returns NULL when it has determined that no solutions exist, as solve does.

Example 19
Use fsolve to apply numerical methods to solve equations

fsolve is similar to solve in that it tries to find a solution to an equation or set of equations.

```
> fsolve(cos(x)=x);
```
$$.7390851332$$

fsolve will behave in one of two ways when it cannot find a solution. It sometimes returns NULL (no printed result), as solve does.

```
> fsolve({x^2+y^2=1,x^3*y+2*x*y=3}, {x,y});
```

For other problems, fsolve will return the input expression, as int or sum do.

```
> fsolve(sin(x)^2+cos(x)^2=2, x);
```
$$fsolve(sin(x)^2 + cos(x)^2 = 2, x)$$

Like solve, fsolve sometimes returns NULL if it determines that the equation has no (real) solutions.

```
> fsolve(x^2+1=0, x);
```

fsolve tries to find all real roots of a univariate polynomial.

```
> fsolve(x^4-x^3-x^2-x-2, x);
```
$$-1.000000000, 2.000000000$$

fsolve cannot find all roots accurately if the problem is very sensitive to rounding errors in arithmetic. "Wilkinson's polynomial", whose exact roots are the integers 1, 2, 3, 4, ..., 20, illustrates this.

```
> p:=                      x^20 -                          210*x^19 +
>              20615*x^18 -                       1256850*x^17 +
>              53327946*x^16 -                   1672280820*x^15 +
>           40171771630*x^14 -                756111184500*x^13 +
>        11310276995381*x^12 -            135585182899530*x^11 +
>      1307535010540395*x^10 -         10142299865511450*x^9  +
>     63030812099294896*x^8  -       311333643161390640*x^7  +
>  1206647803780373360*x^6  -     3599979517947607200*x^5  +
>  8037811822645051776*x^4  -   12870931245150988800*x^3  +
> 13803759753640704000*x^2  -    8752948036761600000*x      +
>  2432902008176640000:
> fsolve(p,x);
1.000000000, 2.000000000, 2.999999998, 3.999999986, 4.999999594,
       5.999999159, 6.999993807, 7.999996971, 8.999939778,
       9.999997186, 11.00001485, 12.00018041, 13.00063387,
       14.00089481, 15.00014113, 16.00066669, 17.00283032,
       18.00358504, 19.00026938, 20.00132486
```

Sometimes increasing the digits of accuracy used in the computation helps. Here we increase Digits from its default of 10 decimal digits, to 20. However, this increases the computation time considerably.

```
> Digits := 20:
> fsolve(p, x);
      1.0000000000000000000, 2.0000000000000000000,
        3.0000000000000000000, 4.0000000000000000000,
        5.0000000000000000000, 6.0000000000000000000,
        7.0000000000000000000, 8.0000000000000000000,
        9.0000000000000000000, 10.000000000000000000,
        11.000000000000000000, 12.000000000000000000,
        13.000000000000000000, 14.000000000000000000,
        15.000000000000000000, 16.000000000000000000,
        17.000000000000000000, 18.000000000000000000,
        19.000000000000000000, 20.000000000000000000
```

fsolve's optional third argument can take on the values given in Table 7.

TABLE 7
The optional third argument to fsolve

When *option* is	fsolve(*expression*, *variable(s)*, *option*) **computes**
complex	One complex-valued root. If *expression* is a polynomial, all complex-valued roots are found.
a..b	If *expression* contains only one variable, fsolve will look for roots only in the open interval (a, b) – all values between, but not including, a and b.
{x = *a..b*, y = *c..d*, ...}	Limit the search when *expression* contains several variables.
maxsols=*n*	Find at most n roots of a polynomial.
fulldigits	By default, fsolve uses fewer than Digits accuracy in its internal calculations when Digits is large, to save time and memory. This default doesn't produce accurate results when the problem is sensitive to minor variations in the numerical parameters of the problem. The fulldigits option causes fsolve to use numbers with Digits precision throughout the computation, though at possibly higher cost.

Example 20
Further examples of fsolve

```
> poly := 23*x^5 + 105*x^4 - 10*x^2 + 17*x:
```

Find all real solutions.	`> fsolve(poly, x);` 0, -4.536168981, -.6371813185

Find all complex solutions.	`> fsolve(poly, x, complex);`
	  0, -4.536168981, -.6371813185, .3040664543 + .4040619058 I,
	    .3040664543 - .4040619058 I

Find all real solutions in the open interval $(-1, 0)$.	`> fsolve(poly, x, x=-1..0);` -.6371813185

Find at most two real solutions.	`> fsolve(poly, x, maxsols=2);` 0, -4.536168981

Find one solution for $x \in [-5, 5]$ by combining options.	`> fsolve(poly, x, x=-5..-.5, maxsols=1);` -.6371813185

Example of a solution constrained in several variables.	`> f := sin(a+b) - log10(a)*b = 0:`
	`> g := a - b^2 = 1:`
	`> fsolve( {f,g}, {a,b}, {a=-1..1,b=-2..2} );`
	    {a = 1.913283552, b = .9556587006}

1.13   Graphing and plotting functions on screen and on paper

The formulae that Maple can compute are a good way of gaining important mathematical insight. So are the numbers and numerical approximations that one can compute through facilities such as Maple's `evalf` command, discussed in Section 1.12.1. In this section, we discuss a third way that Maple can help you establish mathematical understanding — by graphing a function. You can get Maple to do simple graphing through its `plot` command.

1.13.1   Graphing simple functions

`plot( expression, variable = horizontal range )`	Plot of a single function
`plot( expression, horizontal range )`	
`plot( expression, horizontal range, vertical range )`	
`plot( set of expressions, variable = horizontal range )`	Plot of several functions on the same graph

  `plot` will produce a graph of the expression *expression*, that involves one variable *variable* as it varies through the specified *range* of values. If two ranges are specified, the first range will specify the limits of the horizontal axis (the range of the variable *variable*) and the second the limits of the vertical axis. If you wish to display the graphs of several functions on the same plot, use a *set of expressions* for the first argument. Figure 5 illustrates a Maple plot as you would typically see it on the display of a workstation or personal computer. Figure 4 is a plot of two expressions simultaneously. plot's default settings will handle decisions such as plot window size and location, axis labels, and the kind of smooth curves used to connect the points plotted.

Maple has internal default settings for the graphics capabilities of your display. If these settings do not suggest that your display can draw lines and curves, it will plot your function with a "character plot" such as the one drawn in Example 21. While crude, character plots work for almost any kind of display. You may see this behavior even on some displays that can print more than characters — those that support Tektronix or VT100 style graphics, for example. On such displays, you may have to use the plotsetup command before doing your first plot. We discuss this command further in Section 1.13.3.

Example 21
Minimalist plotting: a "character plot" of a single function

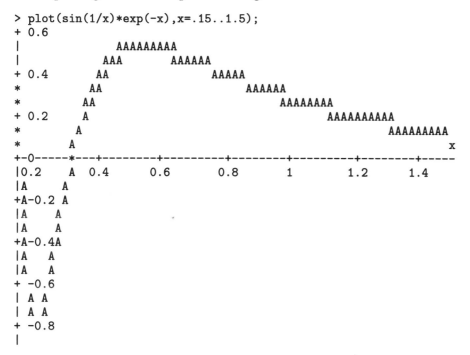

In addition to the kind of plotting shown here, Maple can do:

- parameterized plots,

- three-dimensional (surface) plotting,

- plotting using polar, spherical or cylindrical coordinates,

as well as many more varieties. We discuss plotting further in Chapter 4, starting on page 155.

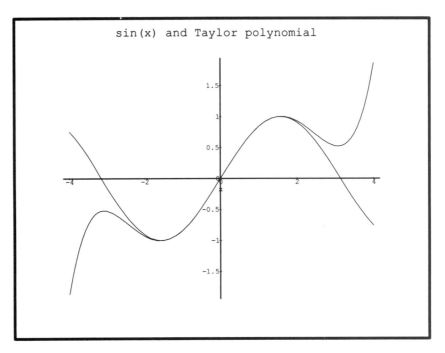

FIGURE 4. Plotting two functions: `plot({sin(x), x-x^3/6+x^5/120},`
`x=-4..4, title=`sin(x) and Taylor polynomial`);`

1.13.2   Printing plots on PostScript, Imagen, and LN03 printers

On some computer systems you can print a plot by selecting the appropriate action from a menu. The DOS, Macintosh, and NeXT versions of Maple work in this way, for example. Consult the Maple documentation specific to your computer system to see if your version of Maple can do this. For other versions of Maple, there are commands to create a file in a graphics format. This file can then be printed or displayed by utility programs you may have on your computer that can process the file format. The format of the file depends on the computer and the printer. After creating the file, you can quit or suspend the Maple session and use the usual commands for printing files for your computer.

There is a four step procedure to print a Maple plot on a PostScript printer.

1. In Maple, enter the command `plotsetup(postscript);`. This sets Maple up for PostScript plotting.

2. In Maple, enter the `plot` command for the graph you wish to print out. Maple will write a PostScript description of the plot onto the file `postscript.out` instead of drawing it on your display.

3. Suspend or quit the Maple session, and use the normal technique for your system for printing PostScript files.[8] To learn how to do this, consult a knowledgeable source for your specific system.

---

[8]On some versions of Maple, you can issue a system command for printing a file from within Maple. You do this by prefacing the command with an exclamation mark. For example, on a Unix system with a PostScript printer named `lw`, the Maple command `!lpr -Plw postscript.out` will print the file.

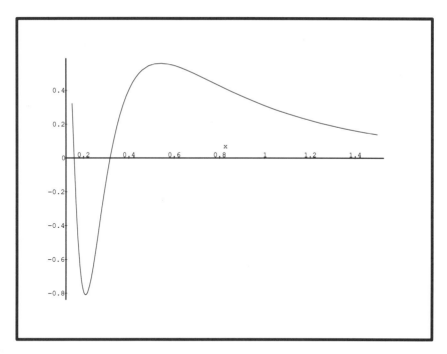

FIGURE 5. A Maple graph of $\sin(1/x)e^{-x}$

4. (a) If you have re-entered your Maple session after printing the plot, you will want the `plot` command to graph on your display again as it did originally, instead of creating graphics files. Use another `plotsetup` command to change things back, so that plots will again appear on your display instead of going to the file. (See Table 8 for a list of displays that `plotsetup` knows about.) For example, `plotsetup(x11);` returns plotting to the display for X Window Systems.

   (b) Alternatively, you can cause the next plot to appear in another file in PostScript format by the command `interface(plotoutput = `*NewFileName*`);`[9]. If you do not change the destination of plot output then the next plot also goes into `postscript.out`. If you had not already copied or renamed what you had recorded in step 2 above it will be lost when the file is rewritten by this action.

   Example 22 and Figure 5 illustrates Maple PostScript plotting. Since PostScript files are "ordinary text", you can edit them with a text editor or word processor. By editing the `scale`, `translate`, or `rotate` specifications in a PostScript format plot file, you can change the printed size or position of the plot. Unfortunately detailed PostScript file manipulation is beyond the scope of this tutorial. If you wish to do this we suggest you make liberal use of local expertise on document processing and desktop publishing.

---

[9]`interface(plotdevice=postscript, plotoutput = `*NewFileName*`);` is an alternative to `plotsetup`. It changes the plotting destination to be the specified file. It also changes the plotting format to PostScript. See the *Maple V Library Reference Manual* or use `?interface` to read more about the `interface` command.

Example 22
How to create a PostScript file for the plot of Figure 5 in `postscript.out`

```
> plotsetup(postscript);
Warning, plotoutput set to postscript.out by default
> plot(sin(1/x)*exp(-x),x=.15..1.5);
```

For Imagen 300 or Digital Equipment Corporation LN03 printers, a similar `plotsetup` command will work. See Table 8.

TABLE 8
`plotsetup` arguments for placing `plot` results in files for printing or further processing

Argument*	What the command `plotsetup(` *argument* `)` does
postscript	A PostScript description of subsequent plots will be written on the file `postscript.out` rather than appearing on your display. The plot will be in the style of Figure 5.
i300	The plot will be written on the file `i300.out` in a format appropriate for Imagen printers.
ln03	The plot will be written on the file `ln03.out` in a format appropriate for Digital Equipment Corporation LN03 printers.
pic	The plot will be written in the file `pic.out` in a format appropriate for input to the Unix program *pic*. See Section 1.13.4.
unix	The plot will be written in the file `unix.out` in a format appropriate for input to the Unix program *plot*. See Section 1.13.4.

*Implementations of Maple for particular computer systems may support an extended set of printers or devices. Check your "Getting started" documentation to see if Maple for your system can generate graphics for other devices.

1.13.3   Plot set-up for graphics terminals and displays

On certain graphics terminals, you should use a `plotsetup` command before plotting your first graph in a Maple session to display plots. `plotsetup` is also necessary after printing plots when you wish to see plotting return to the display.

Table 9 explains the use of `plotsetup` and which terminals it works on.

1.13.4   Creating graphs for other picture processing programs

pic is a program that allows files that describe a picture using *pic*'s own command language to be included in a document processed by the Unix text processor *troff*. `plotsetup(pic);` will instruct Maple to create a *pic* format description in the file `pic.out` of the next graph computed by the `plot` command. Similarly,

TABLE 9
`plotsetup` arguments for graphics terminals and displays

Argument	What the command `plotsetup( argument );` does
`x11`	Prepare Maple for plotting under X Window Systems or DECwindows under VMS. Plots will appear in a separate pop-up window. This is the default for `xmaple`.
`sun`	For SunView use the command `interface(plotdevice=sun);` instead of `plotsetup`.
`mac`	Prepare Maple for plotting in separate pop-up windows on Macintosh computers. (Default for Macintoshes)
`ibm`	For computers running DOS, use `interface(plotdevice=ibm);` instead of `plotsetup` to restore plotting to the display. For information on printing plots on DOS systems, see Appendix C, "Maple under DOS" of the *Maple V Language Reference Manual*.
`regis`	Prepare Maple for plotting on terminals supporting the REGIS graphics protocol.
`tek,vt100`	Prepare for plotting on VT100-style terminals supporting Tektronix 4014 graphics.
`tek,xterm`	Prepare for plotting in an X Window System *xterm* terminal emulation window supporting Tektronix 4014 graphics.
`char`	Prepare for plotting using character plots as in Example 21. This is often the default for terminals that are used only for text entry and display.
`vt100`	Prepare for character plotting using the horizontal line graphics character set of VT100-style terminals. This is especially effective for terminals supporting the VT100's 132 column mode.

`plotsetup(unix);` sets up plotting so that it creates a file `unix.out` suitable for processing by the Unix program *plot*. As with the other plotting modes that create description of graphs in files, you can specify the file to be created by giving the command `interface(plotoutput = NewFileName);`.

1.14 More about syntax errors

Maple has grammatical (syntax) rules for what it considers valid input. For example, Maple requires you to end calculation/commands with semicolons or colons. It won't do anything with what you have entered until the punctuation mark is entered.

When you enter a calculation or command that does not obey Maple's grammatical rules, a *syntax error message* will be printed out.[10] These messages are of the form "`syntax error:`", followed by a pointer to part of the line that you have just entered. The ungrammatical part of the command is somewhere before (usually, but not always, right before), the line fragment.

Example 23
A collection of syntax errors

<table>
<tr>
<td>An example with too many parentheses.</td>
<td>

```
> (a + b))/2 +c;
syntax error:
(a + b))/2 +c;
            ^
```

</td>
</tr>
<tr>
<td>Some words are reserved by Maple for its programming language. For example, the words if and done cannot be used as mathematical symbols or as names of functions.</td>
<td>

```
> if + done;
syntax error:
if + done;
   ^
```

</td>
</tr>
<tr>
<td>Sometimes the error occurs before the line cited. Here, the user forgets to enter a semicolon after 2 + 3.</td>
<td>

```
> 2 + 3
> 2 + 4;
syntax error:
2 + 4;
 ^
```

</td>
</tr>
<tr>
<td>What happens if several lines are invalid? Maple will discard entire lines. In that way, you can get back on track by entering a fresh line with a valid Maple command.</td>
<td>

```
> Sometimes Maple gets sufficiently confused that
syntax error:
Sometimes Maple gets sufficiently confused that
                ^
> it ignores all input on a line.
syntax error:
ently confused that;it ignores all input on a line.
                           ^
> 2 + 2;
```

                                                    4

</td>
</tr>
</table>

As was mentioned in Section 1.9, you must surround Maple "strange names" with backquotes (`` ` ``). Since backquoted names can even include *carriage return*, *newline*, or *line feed* characters, Maple will not consider as an error a line that has a starting `` ` `` but no ending `` ` ``. However, in that situation, it will print out a warning that you are creating a name that is longer than one line.

---

[10]The exact form of error messages may differ slightly from system to system, but should supply essentially the same information as described in this section.

Example 24
What happens when your backquotes don't come in matching pairs

The user forgot to terminate
Funny name with a closing
backquote just before the
semicolon. Note how each
succeeding line of input is
absorbed as part of the name
being formed until a matching
backquote finally is typed.

```
> p := 1 + `Funny name;
Warning: String contains newline character.
Close strings with ` quote.
> p
Warning: String contains newline character.
Close strings with ` quote.
> What happened?
Warning: String contains newline character.
Close strings with ` quote.
> Oh, I forgot the other quote`;

p :=
    1 + Funny name;

p
What happened?
Oh, I forgot the other quote
```

The variable p now has a very
strange value!

1.15   You ask too much! (Run-time errors)

Maple is limited to the realm of what is mathematically possible. You cannot get it to divide by zero, or to compute $\tan(\pi/2)$. It is also limited to what its programming and computing resources allow it to do.

When you tell Maple to do something it is not permitted to do, such as to divide by zero or to create a number or expression that is larger than its limits will permit, it will give you a *run-time error message* beginning with the phrase **Error,** or **Error, (in** *function***)**, where *function* is the name of the Maple function where the difficulty is occurring. Sometimes you will see that the function named in the error message is not the command you specified, but rather a subprocedure whose computation was supposed to be a step in achieving the overall result.

- *ATTENTION*

   Run-time error messages only describe the symptom rather than the cause of an erroneous computation. It is the signal that something was asked for that Maple can't do (e.g. an undefined operation). Such errors are usually corrected by thinking about what you intend to compute, and finding the difference between that and what you said that caused the error.

Example 25
A collection of run-time errors

What happens when you ask
Maple to divide by zero.

```
> 1/0;
Error, division by zero
```

Can't differentiate sin x with
respect to 5, since 5 isn't a
variable .

```
> diff(sin(x),5);
Error, wrong number (or type) of parameters in function diff;
```

The following causes an error
because Maple determines
before computing that the result
could not fit in the computer's
memory. Can you estimate how
many digits long the answer
would be if it were displayed?

```
> 12345667890^9876543210;
Error, integer too large in context
```

Here's another example of what
happens when you try to create
something larger than Maple's
built-in limits permit. The sum
has too many terms – at present
Maple permits objects with
$2^{16} - 1$ (65,535) or fewer terms.

```
> sum(f(i),i=1..2^16);
System error, object too large
```

Some run-time error messages
do not indicate errors by the
user, but rather inadequacies in
Maple's programming. In this
version of Maple, solve has no
techniques for solving the type
of problem given to it.

```
> solve({x^6+3*x+1>2},{x});
Error, (in solve/ineqs) unable to handle these expressions
```

On some computers there are some run-time situations that are so severe that your Maple session "crashes". The example below shows what happens if you assign the formula $x + 1$ as the value of the variable x in a computer running the Berkeley Unix operating system. The session ends abruptly because use of the circular definition of x has exhausted available memory.

Example 26
A catastrophic error

```
> x := x+1;
                    Warning: Recursive definition of name
                                x := x + 1
> x;
/usr/local/bin/maple: 8576 Memory fault - core dumped
```

The message `Recursive definition of a name` serves to warn the user of the potential for disaster. It is one of the few known errors users can do that cause such catastrophic failure. If you encounter persistent and reproducible Maple crashes, be sure to consider this as one of the "prime suspects" in your search for a cause.

1.16   Interrupting a Maple computation

If you decide you don't really want to spend the time and cost of waiting for a Maple computation to finish, it is possible to use the standard interrupt mechanism for your computer system. On some systems, this is a character (for example, Control-Z for VMS, command-. (command-period) on Macintosh or NeXT, or Control-Alt-LeftShift for DOS) or a special key (e.g. Break or Attention). On others, such as the X Window System version, there is a button displayed in the Maple window or a menu item labeled **Interrupt** that you can select with the mouse. What happens when you interrupt Maple depends on the computer system you are using. Typically after interrupting Maple you will see the prompt, and can then continue with another command. On some computers, interrupting Maple ends the Maple session, just as if you had typed `done` or `quit`.

1.17   Printing values: `print` and `lprint`

`print( `*expression₁*`, ..., `*expressionₙ*` )`	Prettyprint expressions
`lprint( `*expression₁*`, ..., `*expressionₙ*` )`	Print expressions in linear format

By default, Maple prints the result of evaluating any expression or assignment you enter. (As we have mentioned previously, ending a command with a colon instead of a semicolon will prevent the printing of the result.) `print` and `lprint` are commands for displaying the values of one or more expressions at once. `print` with several arguments "prettyprints" their values, separated by commas. `lprint` prints out the values that result from evaluating *expression₁*, *expression₂*, and so on. "lprint" stands for "linear print". The results of the command are printed in a linear format like that of Maple input instead of the two-dimensional format used in prettyprinting. Thus, the values will all appear on one line (subject to the line length limitations of your display).

For both `lprint` and `print`, variable names surrounded by left quotes (`) will be printed without them.

Example 27
print and lprint

<table>
<tr><td>Prettyprinting of results is the
default.</td><td>

```
> a := 3/47;
```

$$a := 3/47$$

```
> b := (a+x)^2/x;
```

$$b := \frac{(3/47 + x)^2}{x}$$

</td></tr>
<tr><td>lprint prints results in linear
format.</td><td>

```
> lprint(a,b,a-1);
3/47    (3/47+x)**2/x    -44/47
```

</td></tr>
<tr><td>The same result in prettyprinted
format.</td><td>

```
> print(a,b,a-1);
```

$$3/47, \; \frac{(3/47 + x)^2}{x}, \; -\frac{44}{47}$$

</td></tr>
<tr><td>lprint does not separate items
by commas.</td><td>

```
> lprint(`a's value is`,a);
a's value is    3/47
```

</td></tr>
<tr><td>print does separate items by
commas.</td><td>

```
> print(`a's value is`,a);
               a's value is, 3/47
```

</td></tr>
</table>

1.18   Defining simple functions in Maple

function name := *variable* -> *expression*;	Define a function of one variable
function name := (*var*$_1$, ..., *var*$_n$) -> *expression*;	Define a function of several variables

As we have seen in Section 1.11, Maple has built-in programming for functions commonly used in mathematics, science, and engineering. With Maple, you can evaluate them, differentiate them, solve equations or simplify expressions involving them, etc. You can define your own functions using Maple's "arrow" notation (->). In doing so, you also let Maple know how to evaluate the function when it appears in Maple expressions. The assignment (:=) operation can then associate a function name with the function definition. The name of the function is on the left-hand side of the :=. The function definition (using the arrow notation) is on the right-hand side. For example, f := x -> x^2; is the Maple command to define f as the "squaring function". Once you have given this command, then evaluating expressions such as f(5) or f(y+1) produces the square of f's argument as the value of the expression (25 or $(y + 1)^2$, respectively).

Example 28 illustrates the definition and use of simple functions in Maple, including use of the second form
of the "arrow notation" to define functions of several variables.

Example 28
The arrow notation for defining simple functions in Maple

Define the squaring function,
and use it.

```
> f := y -> y^2;
                                        2
                            f := y -> y
> sin(f(x+1)) * f(f(5))/sqrt(f(Pi+y));
                                    2
                           sin((x + 1) )
                    625    -------------
                              Pi + y
> diff(f(z),z);
                         2 z
```

Define a simple function of
several variables, and use it. The
hypergeometric distribution
function is based on the built-in
binomial function. In the way
we are using it here, it describes
the probability of drawing D
balls from an urn with R red
balls, and B black balls, and
having r of them turn out to be
red.

```
> hyperg := (R,B,D,r) -> binomial(R,r)*binomial(B,D-r) /
>        binomial(R+B,D);
                                binomial(R, r) binomial(B, D - r)
            hyperg := (R,B,D,r) -> ----------------------------------
                                        binomial(R + B, D)
```

Print several values, and their
sum.

```
> hyperg(10,10,2,0), hyperg(10,10,2,1), hyperg(10,10,2,2),
> hyperg(10,10,2,0) + hyperg(10,10,2,1) + hyperg(10,10,2,2);
                             10
                    9/38, ----, 9/38, 1
                             19
```

When you have computed an
expression and wish to make a
function definition out of it, use
the unapply command.

```
> expr1 := tan(t):
> expr2 := diff(tan(t),t):
> f := unapply(expr2,t);
                                        2
                        f := t -> 1 + tan(t)
```

unapply can take additional
arguments which are used as
arguments for the function
definition.

```
> expr3 := subs(t=s*t,expr2):
> g := unapply(expr3,s,t);
                                          2
                    g := (s,t) -> 1 + tan(s t)
```

Chapter 3 describes how more complicated functions can be defined using a different kind of notation (proc/end). If you scan the programming of some of the Maple library functions, you may see an alternative to the arrow notation being used for function definitions: < *expression* | *variable* > is synonymous with *variable* -> *expression*. See the *Maple V Library Reference Manual* or ?-> for further details.

unapply (*expr*,*var*); is a Maple command that creates a function based on the value of *expr* whose argument is named *var*. This command is useful when you have computed an expression *expr* and then wish to use it as a function. Example 28 illustrates the use of this command.

1.19   Automatic simplification

Maple is designed to leave most expressions in the form in which you type them, or the way they are created during a computation. There are exceptions to this philosophy where the Maple designers believed that almost all users would want certain transformations automatically performed — for example, turning 0*x into 0, or turning x*2 and x+x into 2*x. Such transformations of expressions are referred to as *simplification*. We summarize the most noticeable features of Maple's automatic simplification:

1. Arithmetic on sums, products, and powers of integers and rational numbers is done automatically.

2. Greatest common divisors are removed from the numerator and denominator of rational numbers.

3. Syntactically identical factors in the numerator and denominator of a rational expression are removed.

4. Products of terms are rearranged so that the constant term is the first term of the product.

5. Like terms in a sum or product are collected, e.g.$x + x \rightarrow 2 * x$ or $x * x \rightarrow x\hat{}2$.

6. If an expression is a product of a number and a sum of *monomials*,[11] the number is distributed into the sum. For example, $\frac{1}{2}(x + y) \rightarrow x/2 + y/2$.

7. Maple has built-in knowledge about how to simplify the standard mathematical functions for certain arguments. For example, $\cos(\pi/4) \rightarrow \sqrt{2}/2$, $\ln(1) \rightarrow 0$, binomial(n,1) $\rightarrow n$. Transformations that do not eliminate the function, e.g. $\ln(x^2) \rightarrow 2\ln(x)$ are typically *not* automatic. Section 1.20 discusses how to use Maple's simplify command to invoke Maple's built-in knowledge on this second class of transformations.

Example 11 in Chapter 1 demonstrates these automatic simplification rules at work.

When Maple determines that a newly-entered expression is equivalent to an expression already entered into the system, it will be expressed in the same way as before. For example, $x + y + z$, $y + x + z$, $y + z + x$ (as well as the other possible permutations) will all be transformed into one form (probably $x + y + z$, but it varies from session to session and from machine to machine) as part of Maple's automatic simplification. For multivariate expressions, the user can use the built-in functions collect and sort to re-order an expression in terms of a main variable or variables. See Sections 2.3.4 and 2.3.5.

---

[11]A monomial is a product of symbols or functions. For example, if x and y are mathematical symbols, then xy, and $x\,\sin(x^2 + y)\,y^2$ are monomials.

1.20 Simplifying expressions with `simplify`

`simplify( `*expression*` )`	Simplify an expression
`simplify( `*expression*`, `*rule*` )`	Use a specific type of simplification rule

Maple's automatic simplification does not always put an expression into simplest terms. For example, it does not factor or expand polynomial expressions, nor does it remove greatest common divisors from rational functions (ratios of two polynomials). Many transformations can be expensive to compute; they don't always have the effect of simplifying the answer; and often the notion of "simple" is a matter of taste for the individual. Maple leaves these transformations for you to apply as you choose.

`simplify` is the "general purpose" simplification command. It will do the right thing in many, although not all, situations. For example, it always simplifies rational expressions, turning $x^2 - y^2 - (x+y)(x-y)$ into 0. It will also apply all of the standard kinds of simplifications for the functions given in Table 10 below. Through `simplify`, you can have Maple simplify $\sin^2 x + \cos^2 x$ to 1, or $\log x^2$ to $2 \log x$. With the second form of the `simplify` command, you can specify that Maple should perform only a particular simplification *rule*. The *rule* should be a name one chosen from the first column of Table 10.

Example 29
Examples of the simplify command

Automatic simplification leaves this expression alone.

```
> expr := 2*(cos(2*y))^2 + 3*(sin(2*y))^2 +
>      exp(x)^(5/2) * (1+2*x+x^2)^(1/2) + 2^(5/2);
expr :=
                2                  2          5/2             2 1/2
    2 cos(2 y)   + 3 sin(2 y)   + exp(x)    (1 + 2 x + x )
                    1/2
          + 4 2
```

The `simplify` command uses certain transformations to simplify the expression.

```
> simplify(expr);
              2                                         1/2
    - cos(2 y)   + 3 + exp(5/2 x) + exp(5/2 x) x + 4 2
```

The second argument to `simplify` specifies that only one kind of simplification should be performed.

```
> simplify(expr,power);
              2                  2                         2 1/2
    2 cos(2 y)   + 3 sin(2 y)   + exp(5/2 x) (1 + 2 x + x )
                  1/2
          + 4 2
> simplify(expr,trig);
              2                    5/2           5/2        1/2
        - cos(2 y)   + 3 + exp(x)     + exp(x)       x + 4 2
```

TABLE 10
Varieties of transformations available with `simplify`

Variety	Simplification involving	Examples
atsign	Functional operators	`sin @ arcsin` $\rightarrow$ `x -> x`
GAMMA	The Γ function	`GAMMA(n+1)/GAMMA(n)` $\rightarrow$ `n`
hypergeom	Hypergeometric functions	`hypergeom([1],[1],z)` $\rightarrow$ `exp(z)`

power	Powers, exponentials, logarithms	`(a^b)^c` $\rightarrow$ `a^(b*c)`
		`exp(5*ln(x)+1)` $\rightarrow$ `x^5*exp(1)`
		`ln(x*y)` $\rightarrow$ `ln(x) + ln(y)`

radical	Expressions involving radicals (fractional powers)	`(x^2 - 4*x + 4)^(1/4)` $\rightarrow$ `(x - 2)^(1/2)`

RootOf	Expressions including the `RootOf` function	`RootOf(x^2-2=0,x)^2` $\rightarrow$ `2`
		`1/RootOf(x^2-2=0,x)` $\rightarrow$
		    `1/2 * RootOf(x^2-2=0,x)`

sqrt	Square roots or powers of square roots	`16^(3/2)` $\rightarrow$ `64`
		`(10*x^2 + 60 * x + 90)`
		    `^(1/2)` $\rightarrow$ `10^(1/2)*(x+3)`

trig	Powers of trigonometric functions	`1+tan(x)^2` $\rightarrow$ `1/cos(x)^2`
		`cos(2*x)+sin(x)^2` $\rightarrow$ `cos(x)^2`
		`cos(x)^2 + sin(x)^2` $\rightarrow$ `1`

`simplify` always applies the `normal` function as well as these kinds of transformations. The `normal` function is explained in Section 2.3.

This is one of the roots of $x^3 - 12x^2 + 44x - 48 = 0$. You might get this if you were working out the solution by hand using the general formula for solution of a cubic equation.

```
> -1/2*(-64/27)^(1/6)-1/2 * (-1)^(1/3) * (-64/27)^(1/6)
> + 4 + (-3)^(1/2) * 1/2 * ((-64/27)^(1/6)
> - (-1)^(1/3) * (-64/27)^(1/6));
            1/6  5/6              1/3       1/6  5/6
  - 1/54 (-64)    27   - 1/54 (-1)   (-64)    27    + 4
             1/2           1/6  5/6                    1/3
      + 1/2 (-3)   (1/27 (-64)    27   - 1/27 (-1)
          1/6  5/6
      (-64)    27   )
```

simplify simplifies the subexpressions involving radicals (n^{th} roots).

```
> simplify(");
                                        6
```

1.21   Maple's commands for calculus

In this section we discuss how Maple can help you with the basic operations from calculus.

1.21.1   Differentiation

`diff( expression, variable of differentiation )`	Differentiate an expression

`diff` computes the derivative of *expression* with respect to a variable.

Example 30
First derivative with respect to x via `diff`

$$> \text{diff(sin(ln(x^2 + 1) / 3), x);}$$

$$2/3 \; \frac{\cos(1/3 \ln(x^2 + 1)) \, x}{x^2 + 1}$$

Higher-order partial derivatives can be taken by `diff` with a sequence of variables. The following example computes the partial derivative $\partial^2/\partial t \partial s$ of an expression in s and t.

Example 31
Partial derivative with respect to s and t

Compute a second order partial derivative of an expression.

```
> (t^2 -s) / (s^3 -1):
> diff(", t, s);
```

$$- 6 \; \frac{t \, s^2}{(s^3 - 1)^2}$$

Maple also has a *differentiation operator* D that differentiates functions (as opposed to expressions as `diff` does). We discuss them in Section 1.21.3.

1.21.2   Indefinite and definite integration

`int( integrand, variable of integration )`	Indefinite integration
`int( integrand, variable of integration = low .. high )`	Definite integration

Maple will try to calculate a definite or indefinite integral when it is given the `int` command. Yet unlike differentiation, where the computer coding of "how to differentiate" is straightforward, there is no guaranteed success for integration. When Maple fails to solve an integration problem, the request for the calculation, perhaps simplified somewhat, is given as the symbolic "answer".

Example 32
Symbolic indefinite and definite integration with `int`

$\int x^3 \cos x\, dx$.

```
> int(x^3*cos(x), x);
         3             2
        x  sin(x) + 3 x  cos(x) - 6 cos(x) - 6 x sin(x)
```

$\int_a^{2a} x^3 \sqrt{x^2 - a^2}\, dx$.

```
> int( (x^3)*sqrt(x^2-a^2), x=a..2*a);
                        5  1/2
                  14/5 a   3
```

Another indefinite integral:
$\int \frac{x^2}{\sqrt{1-x^2}}\, dx$.

```
> int( (x^2)/sqrt(1-x^2), x);
                         2 1/2
            - 1/2 x (1 - x )    + 1/2 arcsin(x)
```

Other definite integrals.

```
> int( cos(theta)^3, theta=0..Pi/2);
                        2/3
> int( 1/(1+x^2), x=0..infinity);
                        1/2 Pi
```

Sometimes Maple can't compute a closed-form solution.

```
> int( 1/(x+exp(x)), x=0..1);
                    1
                   /
                   |      1
                   |  ---------- dx
                   |  x + exp(x)
                   /
                   0
```

The "answer" is just the prettyprinted version of the original command expression.

```
> lprint(");
int(1/(x+exp(x)),x = 0 .. 1)
```

However, you can use `evalf` to approximate the value of the definite integral.

```
> evalf(");
```

.5163007634

1.21.3 Solving differential equations with `dsolve`

`dsolve` can find closed-form and power series solutions to many first-, second-, or higher-order ordinary differential equations. It also will try to solve systems of equations, either explicitly through the method of Laplace transforms or approximately in terms of power series. There are several forms of the command:

`dsolve(` *equation*, *depvar*(*indvar*) `)`	Solve a differential equation
`dsolve(` *set of an equation and initial conditions*, *depvar*(*indvar*) `)`	Solve an initial value problem
`dsolve(` *set of equations and initial conditions*, *set of variables* `)`	Solve a set of diff. equations
`dsolve(` *set of equations and initial conditions*, *set of variables*, *option* `)`	

Table 11 shows `dsolve`'s options.

TABLE 11
Options used with `dsolve`

When *option* is	`dsolve` (..., *option*); **computes**
`laplace`	A solution using the method of Laplace transforms to solve an equation or system of equations.
`series`	A power-series solution. (The value of the global variable `Order`, with a default value of 6, determines the number of terms.)
`explicit`	A solution using `solve` to force the solution to be returned explicitly in terms of the dependent variable.
`numeric`	Numerical solution of initial-value problems. Example 113 in Section 4.3 contains an example of this option.

Maple's differentiation operator `D` provides another way to write differential equations. In Maple, `D(f)(x)` is the derivative of the function f at x. This notation is more expressive than `diff`, since with it we can express ideas such as `D(f)(0)`, the derivative of f at 0. Higher derivatives may be expressed through `D@@2 (f)`, `D@@3 (f)`, and so forth. Initial conditions involving derivatives must be stated using `D` notation (see Example 33). Section 3.11 discusses further Maple function operators such as `D`.

0

Example 33
Solving differential equations with `dsolve`

A differential equation is an equation involving an unknown function y of an independent variable x. It is necessary to say `diff(y(x),x)` instead of `diff(y,x)` because `diff` needs explicit declaration of y's dependency upon x.

```
> deq := diff(y(x),x)*x^2+y(x)=0;
              / d     \  2
       deq := |---- y(x)| x  + y(x) = 0
              \ dx     /
```

A differential equation solution is requested by including the equation, and the names of the dependent and independent variables.

```
> dsolve(deq, y(x));
              y(x) = exp(1/x) _C1
```

An initial value problem is described by giving a set of equations as the first argument.

```
> dsolve({deq, y(1)=a},y(x));
                    exp(1/x) a
             y(x) = ----------
                      exp(1)
```

Part of `dsolve`'s programming includes Kovacic's algorithm for solving second order linear homogeneous equations with rational function coefficients [Kov86].

```
> dsolve(x^2*diff(y(x),x,x) + 5*diff(y(x),x) = 0,y(x));
     y(x) = _C1 x exp(5/x) - 5 _C1 Ei(5/x) + _C2
```

`dsolve` can't solve all differential equations exactly, of course. It returns "nothing" (the `NULL` value) in that case, similar to `solve`.

```
> dsolve(  diff(f(x),x) + f(x)^5*x = sin(x), f(x));
```

Sometimes the first terms of a power series solution can be found, even if Maple can't find an exact solution.

```
> dsolve( {f(0)=1/2, diff(f(x),x) + f(x)^5*x = sin(x)},
>  f(x), series);
                         31   2    977   4        6
         f(x) = 1/2 + ---- x  - ----- x  + O(x )
                         64       12288
```

Solve a system of differential
equations with initial conditions,
for a set of functions.

```
> sys := { diff(y(x),x) = z(x), diff(z(x),x) = y(x),
      y(0) = 0,                    z(0) = 2 };
                                  d
    sys := {z(0) = 2, y(0) = 0, ---- z(x) = y(x),
                                  dx
          d
         ---- y(x) = z(x)}
          dx
> fcns := {y(x), z(x)};
                          fcns := {y(x), z(x)}
> dsolve(sys, fcns);
      {z(x) = exp(x) + exp(- x), y(x) = exp(x) - exp(- x)}
```

1.21.4   Power series

series(*expression*, *variable* = *expansion point* , *order of truncation*)

Maple can construct any finite number of terms of the Taylor series (and more generally, Laurent series) for an expression. $O(x^ n)$ is tacked on to the end of the series if the terms up to x^{n-1} are not the complete expression. As a rule of thumb, the $O(x^ n)$ usually appears unless the *expression* is a polynomial of degree less than n in the series variable.

The value of Order controls the order of truncation (i.e., the value n above) used during the calculation. The value of Order is initially 6, but you can change it by assignment, just as you would any other variable.

Example 34
Power series calculations

Compute the first few terms of
the Taylor series for $\sin x$
expanded at $x = 0$. The default is
to compute up to $O(x^6)$.

```
> t1 := series(sin(x), x=0);
                         3          5       6
          t1 := x - 1/6 x  + 1/120 x  + O(x )
```

An optional third parameter
specifies the order to override
the default of 6.

```
> t2 := series(ln(x), x=1, 4);
                        2           3            4
  t2 := x - 1 - 1/2 (x - 1)  + 1/3 (x - 1)  + O((x - 1) )
```

The order information in series
results is used when doing
arithmetic on series themselves.
In this example, the result is only
$O(x^7)$ since $t1$ is only $O(x^6)$.

```
> S := series(t1*t1, x=0, 10);
                    2          4          6       7
        S := x  - 1/3 x  + 2/45 x  + O(x )
```

To evaluate a series at a value, compute it, then convert it to a polynomial to remove the order term using the built-in convert function.

```
> P := convert(", polynom);
                        2        4         6
            P := x  - 1/3 x  + 2/45 x
```

Substitute a value for the independent variable.

```
> subs(x=2, P);
                        68
                       ----
                        45
```

Direct evaluation of a series is not allowed unless the order term is absent.

```
> subs(x=2, S);
Error, invalid substitution in series
```

To eliminate high-order terms from a series polynomial, divide out by the lowest-order power you wish to remove, using the built-in function rem.

```
> rem(P,x^6, x);
                    2       4
                   x  - 1/3 x
```

Polynomials have series expansions that are exactly correct if the order of expansion is larger than the degree of the polynomial.

```
> series(x^10 + 1, x=0, 20);
                          10
                     1 + x
> series("*", x=0, 25);
                          10      20
                   1 + 2 x   + x
> series(x^10+1, x=0,5);
                          10
                     1 + O(x  )
> series("*", x=0, 25);
                          10
                     1 + O(x  )
```

Maple has a separate command, asympt for computing asymptotic series. Here we use it to generate Sterling's famous approximation to $n!$.

```
> S1 := simplify(asympt(n!, n, 1));
              / 1/2    1/2  1/2        1  \  n
     S1 := |2    Pi    n    + O(----)|  n  exp(- n)
              |                       1/2 |
              \                      n    /
```

To get rid of the order term of an asymptotic series, we can substitute the "zero function" for the "big-oh function".

```
> simplify(subs(O=0,S1));
                   1/2    1/2  (1/2 + n)
                  2    Pi    n           exp(- n)
```

Maple also can compute Laurent series, and with Puiseaux series.

```
> series(GAMMA(x), x=0,2);
            -1                 2               2         2
           x    - gamma + (1/12 Pi  + 1/2 gamma ) x + O(x )
> series(sqrt(cos(x)), x=0, 4);
                              2      4
                    1 - 1/4 x  + O(x )
```

1.21.5   Limits of real and complex functions

`limit( f, x = a )`	Real bidirectional limit
`limit( f, x = a, option )`	Directional limit

The first form of `limit` computes the limiting value of f, an expression in the variable x, as x approaches the number a. That is, `limit` returns b if $\lim_{x \to a} f = b$, in the "real bidirectional" (standard calculus) sense. As with integration, the programming of `limit` does not always succeed in finding the correct answer. The expression entered is returned when `limit` fails to find an answer.

`limit` assumes that all symbolic unknowns in the expression (other than the limit variable) are *real* and *nonzero*.

The limit point a can be any real number, an expression, `infinity`, or `-infinity`. If the limit point is `infinity` or `-infinity`, a directional limit is taken instead of a bidirectional one. The result of `limit` may be a number, an expression, `infinity`, `-infinity`, `undefined`, or a range $c..d$. The latter occurs when f lacks a limiting value at a or when the limit can't be found by `limit`, but it is known that f's values all lie in the interval $[c, d]$ as x approaches a.

Example 35
Real bidirectional limits

```
> r := (x^2 - 1) / (11*x^2 - 2*x -9):
> limit( sqrt(r), x=1 );
                              1/2  1/2
                      1/10 2     5
> limit( sin(x)/x, x=0 );
                          1
> limit( sin(1/x^2), x=0 );
                      -1 .. 1
> limit( sin(1/x)/x, x=0 );
                      undefined
```

Deriving the definite integral of
sin from 0 to 1 the hard way —
by taking the limit, as $n \to \infty$, of
n rectangles each of width $1/n$.

```
> intsin01 := sum( sin(i/n)*1/n, i=1..n );
                         n + 1                        n + 1
                    sin(-----)              sin(1/n) cos(-----)
                         n                               n
  intsin01 := - 1/2 ---------- + 1/2 -------------------
                        n                    n (cos(1/n) - 1)
                    sin(1/n)          sin(1/n) cos(1/n)
            + 1/2 -------- - 1/2 -----------------
                     n             n (cos(1/n) - 1)
> limit( intsin01, n=infinity );
                       - cos(1) + 1
```

signum is the sign function. (See
Table 5.)

```
> limit( c*x, x=infinity );
                    signum(c) infinity
```

The second form of `limit` computes directional limits. Table 12 explains the options. The `complex` option computes the multidirectional limit. A result of `infinity` here does not mean real positive infinity, but unbounded complex values as the limit point is approached.

TABLE 12
Taking directional limits

When option is	`limit` (*f*, *x=a*, *option*); **computes**
`left`	$\lim_{x \to a^-} f$.
`right`	$\lim_{x \to a^+} f$.
`complex`	A value of `infinity` here means unbounded complex values as the limit point is approached.
`real`	The same result as the real bidirectional limit (i.e., when there is no directional *option*), unless the limit point is infinity. In this situation b is the result if $b = \lim_{x \to -\infty} f = \lim_{x \to +\infty} f$, otherwise the result is `undefined`.

Example 36
Directional limits and varieties of limits at infinity

One-sided limits.

```
> limit(tan(x), x=Pi/2, right);
                            - infinity
> limit(tan(x), x=Pi/2, left);
                             infinity
```

The two-sided real limit doesn't
exist ....

```
> limit(tan(x), x=Pi/2);
                            undefined
```

but the complex limit does.	```> limit(tan(x), x=Pi/2, complex);``` infinity
Limit as x approaches ∞ (positive real infinity)	```> limit(-x, x=infinity);``` - infinity
as x approaches $-\infty$ (negative real infinity)	```> limit(-x, x=-infinity);``` infinity
as x approaches ∞ (bidirectional limit)	```> limit(-x, x=infinity, real);``` undefined
as x approaches complex infinity.	```> limit(-x, x=infinity, complex);``` infinity

1.22   Computing sums

sum(*summand*, *index variable* = *low* .. *high*) sum(*summand*, *index variable*)	Definite summation Indefinite summation

Example 37
Summation with sum

The sum of the first 100 integers.	```> sum(i, i=1..100);``` 5050
The definite symbolic sum $\sum_{i=1}^{n} i^2$.	```> sum(i^2, i=1..n);``` $1/3\ (n + 1)^3 - 1/2\ (n + 1)^2 + 1/6\ n + 1/6$
Another definite symbolic sum $\sum_{j=m}^{n} 2^j$.	```> sum(2^j, j=m..n);``` $2^{(n + 1)} - 2^m$
An indefinite symbolic sum is similar to an indefinite integral – namely, it is the function F such that $F(b + 1) - F(a)$ would yield the sum with index ranging from a to b.	```> sum(1/(i^2),i);``` - Psi(1, i)

As with integration, the returned
value is just the request when no
formula can be found.

```
> sum(1/(1-i),i=0..n);
```

$$\sum_{i=0}^{n} \frac{1}{1-i}$$

```
> lprint(");
sum(1/(1-i),i = 0 .. n)
```

1.23   Solving recurrence relations with rsolve

rsolve(*recurrence equation*, *function(variable)*)	Solve a recurrence relation
rsolve(*set of recurrences and initial conditions*, *function(variable)*)	
rsolve(*set of recurrences and initial conditions*, *set of functions*)	Solve coupled recurrence relations

A *recurrence relation* describes a function's result for an argument n in terms of its values for arguments smaller than n. *Difference equations* are a kind of recurrence relation that express the recurrence using subtraction as the main operation. As with differential equations, in recurrence problems there are usually one or more *initial conditions* to get things started. For example,

$$F(n) = F(n-1) + F(n-2) \quad \text{recurrence relation}$$

$$\left. \begin{array}{l} F(0) = 0 \\ F(1) = 1 \end{array} \right\} \quad \text{initial conditions}$$

define the famous Fibonacci sequence of numbers $F_0 = 0$, $F_1 = 1$, $F_2 = 1$, $F_3 = 2$, $F_4 = 3, \ldots$.

rsolve is Maple's command for solving recurrence problems. In the first form of rsolve, the first argument is an equation expressing the recurrence relation. The second argument describes the function rsolve solves for: the name of the *function* and the name of its argument *variable*. If rsolve is successful in finding a solution, it will return the result. If it is unsuccessful, it returns the input expression as int and dsolve do.

The second form of rsolve is useful when you wish to give different initial conditions than the default. The first argument is then a set consisting of two or more equations — one being the recurrence, the other being the initial conditions. If you have *coupled recurrence relations*, with several recurrence relations and several functions to solve for, use the third form of rsolve. Example 38 illustrates the three forms of rsolve.

rsolve can take a third argument ztrans. This causes it to try to solve the difference equations through the method of Z-transforms. See ?rsolve or the *Maple V Library Reference Manual* for further information.

Example 38
Solving recurrence and difference equations with rsolve

Solve a linear recurrence
$f_n = -3f_{n-1} - 2f_{n-2}$.

```
> rsolve(f(n) = -3*f(n-1) - 2*f(n-2), f);
    (2 f(0) + f(1)) (-1)^n + (- f(0) - f(1)) (-2)^n
```

Find a closed-form solution to
the Fibonacci equation.

```
> Fib_soln:= rsolve( {F(0)=0, F(1)=1, F(n)=F(n-1)+F(n-2)}, F );
    Fib_soln :=
                  1/2              1/2 n        1/2              1/2 n
            1/5 5     (1/2 + 1/2 5   )  - 1/5 5     (1/2 - 1/2 5   )
```

Print out the first 11 numbers in
the series, using the solution. We
use the `simplify` command to
simplify the $\sqrt{5}$'s that appear in
the results. Compare this with
the discussion of efficient ways
of computing Fibonacci numbers
in Section 3.10.

```
> [seq( simplify(subs(n=i, Fib_soln)), i=0..10 )];
              [0, 1, 1, 2, 3, 5, 8, 13, 21, 34, 55]
```

`rsolve` can take a third
argument `ztrans`. This causes it
to try to solve the difference
equations through the method of
Z-transforms.

```
> rsolve( {y(n+1)+f(n)=2*2^n+n, f(n+1)-y(n)=n-2^n+3,
>     y(1)=1, f(5)=6}, {y,f}, ztrans );
                             n
              {y(n) = - 1 + 2 , f(n) = n + 1}
```

1.24 Other commands for solving, and other mathematical functions

Maple has several other solvers, including `isolve`, for finding integer solutions to algebraic equations; `msolve`, for finding solutions to equations modulo p; and `simplex`, `minimize`, and `maximize` for solving optimization problems. Besides its solvers, Maple has hundreds of built-in mathematical functions. Some are used widely. Others are primarily used by specialists in fields such as physics or number theory.

All solvers and built-in functions follow the format of having arguments and returning a value. The *Maple V Library Reference Manual* provides an explanation of every Maple command and function. Explanations are also available through the ? command (see Section 1.6).

To discover more about the vast number of mathematical functions available in Maple, use `?inifcns` to find out more about the set of initially-known functions, `?integer` for integer functions, `?polynom` for functions for manipulating polynomials, `?ratpoly` for manipulating rational functions, `?vector` for vector functions, and `?matrix` for functions using matrices. You can use `?packages` to find out more about the many packages of functions for specific areas in mathematics. In particular, you can use `?numtheory`, `?combinat`, `?stats`, `?orthopoly` to learn more about the special functions available for number theory, combinatorial mathematics, statistics, and orthogonal polynomials, respectively.

Maple also has "computer functions" for manipulation of expressions and Maple data structures. These will be discussed in Chapter 2.

Chapter 2

Less Simple Maple

We have seen how to use Maple as a calculator for algebraic and numerical computations, and for graphics. In this section, we explore how to use Maple in more sophisticated ways.

2.1 A few words to experienced programmers

Maple, unlike computer languages such as Fortran or Pascal, was designed to be used interactively. Simple directions such as "give the name a to the formula $(x+1)^2$", or "compute $\sum_{i=1}^{n} i^3$" are given through one-line commands. Type declarations, BEGIN/END's, and other syntactic paraphernalia are avoided in such "one-step programs". More complicated programs are defined by a Maple function (procedure), as explained in Chapter 3. People who know languages such as Pascal should have little trouble learning how to write simple Maple programs.

Maple is designed specifically for algebraic computation: computations with polynomials, trigonometric functions, equations, etc., can be performed conveniently by the user and efficiently by the system. As a consequence, there are many more symbols with predefined meanings than the typical "general-purpose language". For example, Maple places a meaning on `int`, `diff`, `true`, `false`, `solve`, and `taylor` in addition to `if`, `:=`, and `done`. While you can change the meanings of `int` and `diff` (unlike, say, the meaning of `:=`), you must do so carefully, or the built-in mathematical expertise needing integration or differentiation as a sub-facility will cease to work for you.

Another way that Maple differs from other conventional programming languages is that a variable x in Maple can mean just that — a symbol with no particular, or perhaps unknown, value. Combinations of symbols (such as adding them together) form some of the objects that can be printed out, or assigned as a value to a programming variable. Unlike many general-purpose languages, which were not designed to support interactive algebraic manipulation, it is perfectly reasonable in Maple to have symbols without assigned values.

Maple is fairly rich in built-in data structures. Besides arrays (used to represent vectors and matrices among other things), it also has tables, lists, sets, and sequences, as well as specialized mathematical structures such as sums, products, and equations. Procedures (subroutines) and self-disciplined programming are intended to be used when adding new functionality instead of extensive (and expensive) facilities for modularity or data abstraction. Packages of procedures as explained in Sections 2.13.2 and 3.12 are an example of how modularity can be established in Maple.

Maple's language is explained in this chapter and in Chapter 3. When you use it, you are using the same language that has been used by Maple's designers to build Maple's built-in library — hundreds of thousands of lines of programming. While the system continues to evolve from version to version, the present-day language has been put to the tests of efficiency, convenience and ease of program maintenance. It succeeds well enough to allow a great deal of symbolic, numeric, and graphical computing to get done.

2.2   Programming variables and mathematical symbols

In Maple, symbols (other than symbolic constants such as `I`, `Pi`, `true`) play two roles — one as mathematical variables: "unknowns", and the other as programming variables: labels to refer to results already computed. Having assigned and unassigned variables can, and often does, lead to confusion. However, it is counterbalanced by the infuriating inconvenience that would result if the two varieties of variables weren't syntactically mix-and-match.

A unifying viewpoint is that *all* variables in Maple are programming variables. The ones that are used as mathematical symbols have as their value *their own name*; those used as programming variables have something else assigned as their value. Enclosing a variable's name in right quotes (the apostrophe character ´) gives an expression that is the variable's name, rather than its value (if any). For example, `lprint(´x´)` will print x, rather than the value that may be assigned to x. If x is being used as a mathematical unknown (no assigned value), then `lprint(´x´)` and `lprint(x)` produce the same result: x.

A programming variable can be converted back to a mathematical unknown by assigning it its own name as a value. For example, `x := 5;` makes x a programming variable, `x := ´x´;` then converts x back to a symbol. We discuss this idea further in the section on evaluation, Section 2.4.

2.3   More on simplification: specialized simplification commands

While `simplify` often does what you wish, it would be nearly impossible for it to be programmed to do the right thing in all situations. On occasion you will find it worthwhile to use one or more of Maple's commands that perform specific and specialized transformations that simplify certain kinds of expressions. In this section, we discuss some of the most useful transformations. Table 13 summarizes the commands presented in this section.

2.3.1   Expanding products and functional arguments

`expand` distributes multiplications over additions. For example, applying **expand** to $(x + y)(x - y)$ produces $x^2 + yx - xy - y^2$ which is automatically simplified to $x^2 - y^2$. **expand** is useful in situations where you think that this kind of term-cancellation from turning products into sums leads to a simpler result.

expand also applies multiple-angle rules for trigonometric functions, sum-of-argument rules for the exponential function `exp`, and expansion rules for other functions[1].

---

[1]Two library functions, `expandoff` and `expandon`, can be used to suppress the application of such expansion rules if you do not wish to use them. `expand` also has an optional second argument which controls which expansion rules are applied. See `?expand`, `?expandon`, or the *Maple V Library Reference Manual* for further details.

TABLE 13
Summary of commands in this section

Simplifier	What it does
expand	Distributes multiplications over additions. Also applies multiple-angle rules to trigonometric functions, sum of argument rules for exponentials, etc.
combine	Merges terms of an expression.
normal	Puts an expression in the form numerator over denominator and divides out the greatest common divisor of numerator and denominator.
collect	Organizes an expression with respect to a main variable or variables.
sort	Orders terms in ascending or descending order. Options specify whether ordering is alphabetical, by degree, or according to some other criterion.
convert	Convert from trigonometric, exponential, binomial or other functions into alternative forms. Can convert expressions into partial fractions, continued fractions, radicals, RootOf's and many more forms.
factor	Factor polynomials of one or more variables or terms.

Example 39
Expansion of multiplication over addition, and of cos, exp

$$> \text{expr1} := (x+y+z)^2 - (x+y)^2;$$

$$\text{expr1} := (x + y + z)^2 - (x + y)^2$$

$$> \text{expand(expr1)};$$

$$2\,x\,z + 2\,y\,z + z^2$$

expand also knows about multiple-angle rules for trigonometric and exponential functions, among others.

$$> \text{expr2} := \cos(a+b)-\exp(a+b)+ (n+1)!^2/(n!^2);$$

$$\text{expr2} := \cos(a + b) - \exp(a + b) + \frac{((n + 1)!)^2}{(n!)^2}$$

$$> \text{expand(expr2)};$$

$$\cos(a)\,\cos(b) - \sin(a)\,\sin(b) - \exp(a)\,\exp(b) + (n + 1)^2$$

2.3.2   combine: merging terms of an expression

combine is useful in situations where you want to "un-expand" expansions involving trigonometric functions, exponentials or logarithms. Table 14 describes the varieties of transformations available.

Example 40 illustrates the combine function.

Example 40
Simplifications available through combine

```
> e1 := 2*cos(x/2)^2-1;
```
$$e1 := 2\cos(1/2\,x)^2 - 1$$

Transform e1 to cos(x).
```
> combine(e1, trig);
```
$$\cos(x)$$

```
> e2 := cos(x)^2 + 2*sin(x)^2 + tan(x)^2;
```
$$e2 := \cos(x)^2 + 2\sin(x)^2 + \tan(x)^2$$

Eliminate explicit powers of cos and sin.
```
> combine(e2, trig);
```
$$- 1/2\cos(2\,x) + 3/2 + \tan(x)^2$$

combine(...,exp) will combine products of exponentials in an expression.
```
> e3 := (exp(x))^2*exp(y) + 2*ln(x)-ln(y) + (x^a)^2;
```
$$e3 := \exp(x)^2\,\exp(y) + 2\ln(x) - \ln(y) + (x^a)^2$$
```
> combine(e3, exp);
```
$$\exp(2\,x + y) + 2\ln(x) - \ln(y) + (x^a)^2$$

combine(...,ln) will combine sums of logarithms in an expression.
```
> combine(e3, ln);
```
$$\exp(x)^2\,\exp(y) + (x^a)^2 + \ln(\frac{x^2}{y})$$

combine(...,power) will combine powers of powers and products of powers.
```
> combine(e3, power);
```
$$\exp(2\,x + y) + 2\ln(x) - \ln(y) + x^{(2\,a)}$$

TABLE 14
How combine simplifies expressions

combine **phrase**	**What it does**
combine(*expr*, trig)	Eliminates products and powers of cos, sin, cosh and sinh in *expr* using "multiple-angle" rules such as : $$\sin(a)\cos(b) \rightarrow \sin\tfrac{a+b}{2} + \sin\tfrac{a-b}{2}$$ $$\sinh(a)\sinh(b) \rightarrow \cosh\tfrac{a+b}{2} - \cosh\tfrac{a-b}{2}$$
combine(*expr*, exp)	Simplifies products and powers of exponents in *expr* using the rules: $$e^x e^y \rightarrow e^{x+y} \qquad e^{(x^a)} \rightarrow e^{xa} \qquad e^{x+n\ln(y)} \rightarrow e^x y^n$$
combine(*expr*, ln)	Simplifies sums of logarithms and products of logs and non-log expressions in *expr* using the rules: $$a\ln(x) \rightarrow \ln(x^a) \qquad \ln(x) + \ln(y) \rightarrow \ln(xy)$$
combine(*expr*, power)	Simplifies expressions containing products of powers (including exponentials) using the rules: $$x^y x^z \rightarrow x^{y+z} \qquad (x^y)^z \rightarrow x^{yz}$$ $$e^x e^y \rightarrow e^{x+y} \qquad (e^x)^y \rightarrow e^{xy}$$ $$\sqrt{-a} \rightarrow i\sqrt{a}$$ for arbitrary x, y, z, and integer $a > 1$.
combine(*expr*, Psi)	Simplifies *expr* with respect to sums of $\psi^{(n)}$ functions using: $$\psi^{(n)}(z+1) \rightarrow \psi^{(n)}(z) + (-1)^n\, n!\, z^{-n-1}$$ $$\psi^{(n)}(1-z) \rightarrow (-1)^{-n}\left(\psi^{(n)}(z) + \tfrac{d^n}{dz^n}\cot\pi z\right)$$ for positive integer n.
combine(*expr*)	Combines sums of terms in *expr* involving the inert operators Int, Sum, and Limit: $$c_1 \lim_{x\to a} f(x) + c_2 \lim_{x\to a} g(x) \rightarrow \lim_{x\to a} c_1 f(x) + c_2 g(x)$$ $$c_1 \int_a^b f(x)dx + c_2 \int_a^b g(x)dx \rightarrow \int_a^b c_1 f(x) + c_2 g(x)dx$$ $$c_1 \sum_{i=a}^b f(i) + c_2 \sum_{i=a}^b g(i) \rightarrow \sum_{i=a}^b c_1 f(i) + c_2 g(i)$$ for an arbitrary variable x, c_1 and c_2 arbitrary expressions independent of x, and $f(x)$, $g(x)$ arbitrary expressions involving x.

combine will work with lists or sets of arguments. Here, we simplify a list of two expressions using several transformations.

```
> combine([e1,e3],{exp,ln,trig});
```

$$\left[\cos(x),\ \exp(2\ x + y) + (x^a)^2 + \ln\left(\frac{x^2}{y}\right)\right]$$

2.3.3   normal: simplifying rational functions with finesse

There may be situations where the factored result is more succinct than the expanded result, such as $(x+1)^{100}$. Using expand on such an expression would produce a result 101 terms long. Not only is the expanded result difficult to comprehend, but it also takes up much more space in the computer than the factored result. The normal function is designed to be an alternative to expansion. It performs simplification while trying to avoid expansion as much as possible.

The general form of a "normalized" expression combines everything into one rational expression, multiplied by a rational constant. The numerator and denominator will be *relatively prime*. The rational constant will be such that the numerator and denominator in the rational expression will have *integer content* 1.[2]

Example 41
normal can provide a kind of simplification of rational expressions

normal can be used to simplify rational expressions, possibly without expanding numerators and denominators.

```
> (2*x+4*y)*(x^2-y^2)/(x+y)^2;
```

$$\frac{(2\ x + 4\ y)\ (x^2 - y^2)}{(x + y)^2}$$

```
> normal(");
```

$$-2\ \frac{(x + 2\ y)\ (y - x)}{x + y}$$

Normal form ensures that a rational expression equivalent to zero is normalized to zero.

```
> normal( (x^2-y^2) - (x+y)*(x-y) );
                       0
```

---

[2]Two expressions are relatively prime if they have no common factors (other than 1). The *integer content* of a polynomial is the greatest common divisor of the polynomial's integer coefficients.

Normal form is not always the
"most succinct" form, however.

```
> 2*x/(x+y)^3 + x/(x+y)^2 + y/(x+y) + y/(x+y)^0;
            x          x         y
       2 -------- + -------- + ----- + y
            3           2      x + y
         (x + y)     (x + y)
> normal(");
          2              2       2   3       3     2 2          3
    (2 x + x  + x y + y x  + 2 x y  + y  + y x  + 3 y  x  + 3 x y
          4   /          3
       + y ) / (x + y)
            /
```

2.3.4 collect: organizing an expression with respect to a main variable

collect(*expression*, *variable*)	Collect expression with respect to *variable*
collect(*expression*, *expr₂*)	Collect expression with respect to *expr*$_2$
collect(*expression*, *varlist*, *form*, *coefficient simplifier*)	Collect expression and simplify coefficients

collect is the command to use when you wish to structure an expression with respect to a main variable or variables — written as a sum of terms, where each term is a power of a main variable or expression times an associated coefficient. There are two possibilities for the *form* of the result — either recursive or distributed.

With the recursive form, the result is structured hierarchically. It is cast into the form of being a polynomial in the first variable in *varlist*, whose coefficients are polynomials in the second variable in *varlist*, and so forth for the rest of the variables in *varlist*.

With the distributed form, the result is written like an expanded polynomial — a sum of terms. Each term is the product of powers of the *varlist* variables and a coefficient (not involving the variables in *varlist*). This form is useful when you wish to treat several variables as "main" without giving priority to one of them.

The optional fourth argument *coefficient simplifier* to collect should, if given, be the name of a simplification procedure to be applied to the coefficients of each term of the result. For example, if expand is the fourth argument, then it will be applied to each coefficient. Thus it is possible to organize coefficients differently than the hierarchical and recursive form collect imposes upon the pieces of the expression involving the variables in *varlist*.

Example 42 illustrates the ways collect can be used.

Example 42
Combining terms with `collect`

Collect the terms of `expr` with `x` as a primary variable, then `y`, then `z`. This is the recursive form of `collect`.

```
> expr:=r^2*x^2-2*x^2*r+x^2+8*z*y*r^2-2*z*y*r*s
> -6*z*y*s^2+t^2*z^2+3*r^2*s*y*t*x-r*x*s*y
> +s^2*y^2+r*s*y^2+r^3*z*x-r*z*x*t^2;
```

$$expr := r^2 x^2 - 2 x^2 r + x^2 + 8 z y r^2 - 2 z y r s$$
$$- 6 z y s^2 + t^2 z^2 + 3 r^2 s y t x - r x s y$$
$$+ s^2 y^2 + r s y^2 + r^3 z x - r z x t^2$$

```
> collect(expr, [x,y,z]);
```

$$(- 2 r + 1 + r^2) x^2 + ((- r s + 3 r^2 s t) y$$
$$+ (- r t^2 + r^3) z) x + (s^2 + r s) y^2$$
$$+ (- 2 r s - 6 s^2 + 8 r^2) z y + t^2 z^2$$

Using the distributed form, it's easier to see things such as the coefficient of `x*z`.

```
> collect(expr, [x,y,z], distributed);
```

$$(- 2 r + 1 + r^2) x^2 + (s^2 + r s) y^2 + t^2 z^2$$
$$+ (- 2 r s - 6 s^2 + 8 r^2) z y$$
$$+ (- r s + 3 r^2 s t) x y + (- r t^2 + r^3) x z$$

By specifying `factor` as the fourth option, our coefficients will appear in factored form.

```
> collect(expr,[x,y,z], distributed, factor);
```

$$(r - 1)^2 x^2 + s (s + r) y^2 + t^2 z^2$$
$$+ 2 (4 r + 3 s) (r - s) y z$$
$$+ r s (- 1 + 3 r t) x y$$
$$+ r (r - t) (r + t) x z$$

2.3.5   Ordering terms with `sort`

`sort( `*expression*` )`	Alphabetize and sort terms by exponent
`sort( `*expression*`, `*variable list*`, plex )`	Order an expression lexicographically
`sort( `*expression*`, `*variable list*`, tdeg )`	Order an expression using total degree ordering
`sort( `*list*` )`	Sort a list of expressions in ascending order
`sort( `*list*`, `*ordering*` )`	Sort a list using a specified ordering

Maple orders terms consistently. If an expression is printed as $x + y + z$ the first time in a session, then it will continue to do so within the session. However, the ordering Maple finds expedient and efficient to use may not correspond to the ordering you may wish to see in a result. For example, Maple does not order all the terms of a polynomial involving a single variable. If you wish this, sort (*expression*) will order the terms in descending order. Maple will continue to use the rearranged ordering for an expression later within the Maple session. sort (*expression*) will also place sums or products of variables in simple alphabetical order.

For expressions involving powers of several variables (or several kinds of subexpressions), sort takes three arguments. When you conceive of an expression as having a main variable, and wish to order the terms of the expression giving priority to those which have the highest powers of the main variable, sort (*expression*, *variable list*, plex) will order the terms in that way. *variable list* should be a list of variables with the main variable first. To break ties between terms having the same power of the main variable, the power of the next variable in *variable list* is used. If the tie persists, then the next variable is used, and so forth. This is known as *lexicographical ordering* because it is similar to the rules for ordering words alphabetically.

Another way to organize the terms of an expression is according to "total degree". The "total degree" of a product of powers of variables is the sum of all the exponents. (The total degree of $x^5 y^2 z^3$ is $5+2+3 = 10$, for example.) Placing the terms of an expression in order of their total degree is known as *total degree* ordering. sort(*expression*, *variable list*, tdeg) will sort using total degree ordering. In order to break ties between two terms with the same total degree, the lexicographical ordering based on the variable ordering of *variable list* is used.

If a list is given as an argument to sort, then the elements of the list are sorted in ascending order. An optional argument can specify the type of ordering to use. This can be a keyword for certain predefined orderings that Maple understands or it can be a user-defined comparison function. See the *Maple V Language Reference Manual* and the *Maple V Library Reference Manual* or ?sort for further details on *ordering*.

Example 43 illustrates uses for the sort function. sort can also be used to sort data structures such as lists or arrays. For more information, consult the *Maple V Library Reference Manual* or ?sort.

Example 43
Ordering terms with sort

A particular ordering, once established within a Maple session, is used consistently anywhere the expression occurs, or when it occurs as a complete subexpression.

```
> a := y+z+x;
                         a := y + z + x
> b := x^3 + y + sin(x+y+z)+
> factor(x^2 + 2*x*y + 2*x*z + y^2 + 2*y*z + z^2) + z;
          3                                  2
   b := x  + y + sin(y + z + x) + (y + z + x)  + z
```

The default for univariate polynomials does not necessarily produce an ascending or descending ordering of terms.

```
> c := expand((x^2+x+1)^5);
                  3        2       4        6        5
   c := 1 + 5 x + 30 x  + 15 x  + 45 x  + 45 x  + 51 x
            8         7       10      9
      + 15 x  + 30 x   + x   + 5 x
```

sort arranges terms in descending order.

```
> sort(c);
        10        9        8        7        6        5        4
       x   + 5 x  + 15 x  + 30 x  + 45 x  + 51 x  + 45 x
               3        2
         + 30 x  + 15 x  + 5 x + 1
```

sort will alphabetize a list of strings/names.

```
> sort([z,u,w,s,u,v,w,y,t]);
                  [s, t, u, u, v, w, w, y, z]
```

sort orders other objects according to internal machine locations.

```
> sort(t*s*u);
                             t u s
> sort(s+t+u);
                           t + u + s
```

The ordering induced by sort continues to be used subsequently in the session.

```
> sin(t+s+u);
                          sin(t + u + s)
```

An example of lexicographical ordering with x as the main variable.

```
> e := 1+x^4*y*z+y^5*z^2*x+x*y^2*z^3;
                       4          5  2           2  3
              e := 1 + x  y z + y  z  x + x y  z
> sort(e, [x,y,z], plex);
                 4          5  2       2  3
                x  y z + x y  z  + x y  z  + 1
```

In this ordering by total degree, we use the lexicographical ordering [x,y,z] to break ties.

```
> sort(e, [x,y,z], tdeg);
                  5  2      4           2  3
                 x y  z  + x  y z + x y  z  + 1
```

Complete subexpressions can be used to order terms as well.

```
> expand(cos(x+y)-sin(x+y)^2);
                                                2       2
      cos(x) cos(y) - sin(x) sin(y) - sin(x)  cos(y)
                                               2       2
              - 2 sin(x) cos(y) cos(x) sin(y) - cos(x)  sin(y)
> sort(", [cos(x),sin(x),cos(y),sin(y)], tdeg);
               2       2
      - cos(x)  sin(y)  - 2 cos(x) sin(x) cos(y) sin(y)
               2       2
              - sin(x)  cos(y)  + cos(x) cos(y) - sin(x) sin(y)
> sort(", [cos(x),sin(x),cos(y),sin(y)], plex);
           2       2
  - cos(x)  sin(y)  - 2 cos(x) sin(x) cos(y) sin(y)
                               2       2
          + cos(x) cos(y) - sin(x)  cos(y)  - sin(x) sin(y)
```

2.3.6   converting between functional forms

convert(*expression*, *form*)	Convert to another form

The `convert` function can be used to simplify certain kinds of trigonometric expressions involving functions such as sin, cosh, arctan, etc. It can also be used to convert the binomial function into factorials. Table 15 illustrates the various options that can be used for simplification.

`convert` also has options for transforming expressions into continued fraction form (`confrac`), partial fraction form (`parfrac`), and octal base (`octal`). It can also convert RootOfs into radicals and vice versa (`radical` and `RootOf`). See the *Maple V Library Reference Manual* or `?convert` for further details.

`convert` is also used for data structure conversions. This is discussed further in Section 2.11.

TABLE 15
Varieties of transformations available with `convert`(*expression*, *option*)

Option	Action
exp	Converts all trigonometric functions into equivalent forms involving exponentials.
ln	Converts arctrig functions to forms involving only logarithms.
expln	Combination of `exp` and `ln` options.
expsincos	Converts all trigonometric functions to forms involving sin and cos, and all hyperbolic trigonometric functions to forms involving exp.
sincos	Convert trig functions to forms involving only sin, cos, sinh, and cosh.
tan	Converts all trig functions to forms only involving tan.
trig	Converts all exponentials to forms involving sin, cos, sinh and cosh via Euler's rules.
factorial	Converts GAMMAs, binomial and multinomial* coefficients to factorials.
GAMMA	Converts factorials, binomials and multinomial expressions to use the Γ function.
hypergeom	Converts expressions involving sum or Sum[†] into those involving hypergeometric functions (`hypergeom`).

*Multinomial coefficients arise from using the `combinat` package for combinatorics computations. See Section 2.13 for a general discussion of packages and the *Maple V Library Reference Manual* or `?combinat[multinomial]` for more information.

[†]Sum is the function to use instead of the sum of Section 1.22 when you wish to describe a sum, but don't wish Maple to figure out automatically a formula for it as it does with sum. Maple has other *inert operators* similar to Sum: Int, Gcd, etc. They are also used in situations where you do wish to refer to the formal expression without having Maple evaluate it, to be able to manipulate and simplify the expression with the sum, integral, etc. in it.

Example 44
Conversion from one functional form to another with `convert`

<table>
<tr>
<td valign="top">A trigonometric simplification problem.</td>
<td>

```
> expr := cos(a)^2*cos(b)^2 - 2*cos(a)*cos(b)*sin(a)*sin(b) +
>         sin(a)^2*sin(b)^2;
expr :=
           2       2
     cos(a)  cos(b)  - 2 cos(a) cos(b) sin(a) sin(b)
                2       2
         + sin(a)  sin(b)
```
</td>
</tr>
<tr>
<td valign="top">`simplify` doesn't do much with this.</td>
<td>

```
> simplify(expr);
             2       2
   2 cos(a)  cos(b)  - 2 cos(a) cos(b) sin(a) sin(b) + 1
               2          2
         - cos(b)  - cos(a)
```
</td>
</tr>
<tr>
<td valign="top">Simplify through an indirect route.</td>
<td>

```
> combine(expr, trig);
                    1/2 cos(2 a + 2 b) + 1/2
> convert(", tan);
                                  2
                       1 - tan(a + b)
                   1/2 --------------- + 1/2
                                  2
                       1 + tan(a + b)
> simplify(");
                                  2
                       cos(a + b)
```
</td>
</tr>
<tr>
<td valign="top">We saw the formula for the hypergeometric probability distribution in Example 28. We use the `alias` command to use C as an abbreviation for Maple's `binomial` function.</td>
<td>

```
> alias( C = binomial );
                            I, C
> hyperg := (R,B,D,r) -> C(R,r) * C(B,D-r) / C(R+B,D);
                            C(R, r) C(B, D - r)
              hyperg := (R,B,D,r) -> -------------------
                                        C(R + B, D)
```
</td>
</tr>
<tr>
<td valign="top">Another approach to deriving the hypergeometric probability distribution leads to a different formula.</td>
<td>

```
> hyperg1 := (R,B,D,r) -> C(D,r) * C(R+B-D,R-r) / C(R+B,R);
                          C(D, r) C(R + B - D, R - r)
            hyperg1 := (R,B,D,r) -> --------------------------
                                          C(R + B, R)
```
</td>
</tr>
<tr>
<td valign="top">How do we prove that they are the same? Let's define a function that expresses the difference between the two formulae. It should be identically zero (zero symbolically).</td>
<td>

```
> ShouldBeZero := (R,B,D,r) ->
>               (hyperg(R,B,D,r) - hyperg1(R,B,D,r));
    ShouldBeZero :=
        (R,B,D,r) -> hyperg(R, B, D, r) - hyperg1(R, B, D, r)
```
</td>
</tr>
</table>

Let's evaluate at a few places to see if it's plausible that they are equal. This can also be a useful check on whether you've entered or computed something correctly.

```
> ShouldBeZero(10, 10, 5, 1);
                              0
> ShouldBeZero(25, 16, 6, 2);
                              0
```

If this simplifies to zero for arbitrary symbols, then we've shown the two are equal for any possible combination of values.

```
> simplify( ShouldBeZero(R,B,D,r) );
         (C(R, r) C(B, D - r) C(R + B, R)
              - C(D, r) C(R + B - D, R - r) C(R + B, D))/
         (C(R + B, D) C(R + B, R))
```

Conversion of the binomial expressions into factorials achieves the desired result.

```
> convert(", factorial);
                              0
```

Another example: simplifying a recurrence relation via rsolve.

```
> rsolve( T(n) = T(n/2)+n^3, T(n) );
                            /ln(n)      \
                            |----- + 1|
                            \ln(2)     /
                    1/7  8                  - 8/7
```

simplify doesn't do much with this.

```
> simplify(");
                              /ln(n) + ln(2)\
                              |-------------|
                              \    ln(2)    /
                    1/7  8                      - 8/7
```

But we can get further simplification to occur by converting powers to exponentials.

```
> convert(", exp);
                          ln(8) (ln(n) + ln(2))
              1/7 exp(---------------------) - 8/7
                                ln(2)
```

simplify can handle this.

```
> simplify(");
                               3
                          8/7 n  - 8/7
```

2.3.7 Factoring polynomial expressions

`factor( polynomial )`	Factor a polynomial over the rational numbers
`factor( polynomial, field )`	Factor a polynomial over an algebraic extension field

`factor`(*polynomial*) will perform a complete factorization of polynomials in one or more variables over the rational numbers. The second form of the command can be used to specify that a coefficient domain other than the rational numbers should be used for factoring. While modern algorithms have led to a reasonably efficient factorization algorithm, the process of factorization is a relatively expensive operation compared with most of the other simplifying routines, so it should be used conservatively.

Example 45
Factorization of polynomials over the integers via `factor`

<table>
<tr>
<td>Example of univariate
factorization.</td>
<td>

```
> p := x^3 + 3*x^2 + 3*x + 1;
                          3       2
                 p := x  + 3 x  + 3 x + 1
> diff(p, x);
                        2
                    3 x  + 6 x + 3
> factor(");
                              2
                    3 (x + 1)
```

</td>
</tr>
<tr>
<td>Compare the factored form of
the derivative to the derivative
of factored original. They should
be the same!</td>
<td>

```
> factor(p);
                              3
                    (x + 1)
> diff(", x);
                              2
                    3 (x + 1)
```

</td>
</tr>
<tr>
<td>Factoring over the rational
numbers means that only
rational numbers, not complex
numbers, appear in the factors.
The polynomial q is "irreducible
over the rational numbers".</td>
<td>

```
> q := x^2 + 9/4;
                           2
                   q := x  + 9/4
> factor(q);
                       2
                      x  + 9/4
```

</td>
</tr>
<tr>
<td>An example of multivariate
factorization.</td>
<td>

```
> x^4*sin(y)^4 + x^4*sin(y)^3 + x^2 * sin(y) + x^2;
        4      4    4      3    2              2
       x  sin(y)  + x  sin(y)  + x  sin(y) + x
> factor(");
              2                2          3
             x  (sin(y) + 1) (x  sin(y)  + 1)
```

</td>
</tr>
</table>

Using the second form of the factor command, we specify a different domain for the result's coefficients: complex numbers whose real and imaginary parts are rational numbers. Our example polynomial has non-trivial factors in this domain.

```
> factor(q, I);
                    (x + 3/2 I) (x - 3/2 I)
```

Factors whose coefficients are a rational multiple of $\sqrt{2}$. We use the alias command to provide a convenient nickname for the algebraic number RootOf(x^2 +2). alias is an advance feature discussed further in Section 2.14.

```
> alias( alpha = RootOf(z^2 +2) );
                    I, alpha
> factor(x^2+1/8, alpha);
              (x - 1/4 alpha) (x + 1/4 alpha)
```

2.4   Full and delayed evaluation

Consider the sequence of assignments in Example 46. When x was first used as a programming variable, it had the value y. Yet "5" is the value printed out for x on the last line of the example. Whenever Maple does a calculation, the symbols in the expression are checked to see if they are being used as programming variables (that is, whether any of them have been assigned values). If that is the case, it substitutes the assigned value for the symbol in the expression. It then checks to see if there are more value/programming variable substitutions possible, and does them as well. The result, after there are no more substitutions possible, is called *full evaluation*.

Example 46
"Full evaluation" follows the chain of assignments

```
> x := y;
                                        x := y
> y := z;
                                        y := z
> z := 5;
                                        z := 5
> x;
                                          5
```

● *ATTENTION*

> A problem many users of Maple encounter is due to full evaluation. It occurs whenever they use *i* as a programming variable, and then as the index of summation in a `sum` computation. The next example illustrates this problem. It is a good practice to avoid using indexing variables for other purposes.

Example 47
A problem due to full evaluation

Use i as a label for a constant in a calculation.

```
> x := 3:     i := 4:
> x^i!;
                                                    282429536481
```

Now think of i as a mathematical symbol.

```
> sum(i^2,i=1..n);
Error, (in sum) summation variable previously assigned,
                   second argument evaluates to, 4 = 1 .. n
```

i has a value left over from before! It can't be a mathematical symbol.

```
> i;
                                                    4
```

What was the error? i was first used as a programming variable, that is, as a label for a computed result. When the `sum` was computed, Maple first applied full evaluation to the arguments of the `sum` procedure. Thus, `sum(i^2,i=1..n)` turned into `sum(16, 4=1..n)` before the `sum` procedure was invoked. The error message was generated because "4" is not a proper summation index. We explain how to resolve this common situation in Example 50 in Section 2.5, "Quotation and Unevaluation".

While it might be argued that `sum` should be designed not to use "full evaluation", the designers of Maple have opted for consistency, since a general rule is easier to remember across many functions than to keep track of an assortment of cases. Thus, the policy of "full evaluation" has only a few exceptions:

- The `assigned` and `evaln` functions only evaluate their arguments to the point where they become names (i.e. a single variable) and no further. See the *Maple V Library Reference Manual* or `help` for further details about these functions.

- Maple procedures, tables, and arrays have a special kind of evaluation called *last name evaluation*. The result of evaluation is not the procedure or array itself, but the programming variable that is assigned the procedure, table, or array as its value. This will be illustrated further in the Sections 2.10.1 and 3.5, "Tables" and "Simple Maple procedures".

- The concatenation operator (.), discussed in Section 2.7, does not evaluate the first item in a concatenation expression. But it is a peculiarity in other ways, since concatenation expressions are one of the few expressions permitted on the left-hand side of an assignment operator – for example `a.i := i;`.

- The local variables of a procedure use what is called "one level evaluation". If a local variable has been assigned, then the result of evaluation is the most recently assigned value. Otherwise, it is the unassigned name. This is discussed in Section 3.6, "Maple Procedures".

- Several functions have special evaluation rules. They are `eval`, `evaln`, `evalhf`, `seq`, `assigned`, and `userinfo`. See the *Maple V Language Reference Manual* for details.

2.5   Quotation and unevaluation

Enclosing a variable in right quotes (´) is the way to specify the *name* of a programming variable instead of its *value*. Quoting a variable or an expression prevents evaluation of the variables enclosed by the quotes.[3] However, quoting does not prevent simple arithmetic or most of the rest of the operations of the automatic simplifier (see Section 1.19), e.g. $1 - 1 \to 0$, $y/y \to 1$. The result of "fully evaluating" a quoted expression is to strip off one layer of quote marks.

Example 48
The effect of evaluation, levels of quotation

x used as a programming variable.	`>  x := 3;`	`x := 3`

Quoting x gives a polynomial in x instead of 4.	`>  x + 1;`
	
	 `>  'x' + 1;`
		`4`
		
		 `x + 1`

Full evaluation of a quoted expression or variable strips off one level of quotation.	`>  '''x'+1'';`
	
	 `>  ";`
	
	 `>  ";`
	
	 `>  ";`
		`''x' + 1'`
		
		 `'x' + 1`
		
		 `x + 1`
		
		 `4`

Quoting an expression delays substitution of the values of the programming variables and functions within it, but does not prevent arithmetic.	`>  '(3 + x - 3 + 3*y/y) mod x';`	`(3 + x) mod x`

---

[3]See the *Maple V Language Reference Manual* Section 3.2.11, "Unevaluated Expressions".

A special case of "unevaluation" arises when you wish to remove the assigned value from a programming variable — that is, to restore it to being a mathematical symbol. To do this, one assigns the variable its name, e.g. `x := ´x´;`.

Example 49
Assigning and unassigning variables

Assign x the value 3, and print out the value of x+1 .

```
> x := 3: x + 1;
```
$$4$$

Unassign x – restore it to being a mathematical symbol.

```
> x := 'x';
```
$$x := x$$

Print out the value of x+1 .

```
> x+1;
```
$$x + 1$$

- *ATTENTION*

To resolve the situation with sum and i mentioned in Example 47 (Section 2.4), one could remove any assigned value from i, or quote i wherever it appears in the sum command. As a precaution, you can quote, or unassign, a variable in an expression or command if you wish to use it as a mathematical unknown with no value, but aren't sure whether it has previously been assigned a value in the Maple session.

Example 50
Resolving the problem with i and sum.

Use i as a label for a constant in a calculation.

```
> i := 4:
```

4 can't be used as a variable of summation because it has a value.

```
> sum(i^2,i=1..n);
Error, (in sum) summation variable previously assigned,
                     second argument evaluates to, 4 = 1 .. n
```

To proceed, remove the assignment of i before summing.

```
> i := 'i';
```
$$i := i$$
```
> sum(i^2,i=1..n);
```
$$1/3 (n + 1)^3 - 1/2 (n + 1)^2 + 1/6 n + 1/6$$

2.6   Using quoted variables as function arguments

Maple functions follow the mathematical tradition that functions are single valued. Sometimes Maple functions bend this notion a bit and return a set of expressions, but since a set is a valid Maple entity this is still one "value".

Yet sometimes it is desirable to return several results through separate channels. Quoted variables are often used as additional arguments to Maple functions to export extra information from them. Certain Maple functions follow a programming rule: optional extra arguments are for exporting extra information. These optional arguments must be quoted or unassigned variables.

During the course of computing the function's value, these export variables are assigned additional information that the function computes and exports.

`divide` is a built-in Maple function that can export extra information through additional arguments. `divide(p,q)` computes the quotient of two polynomials p and q that are given as the arguments to the procedure. The value returned by `divide` is `true` if q divides p, and `false` otherwise. During the course of determining whether q divides p, the quotient p/q is formed, but there is no way to return it as the value of the function since either `true` or `false` is being returned through that channel. However, you can give an optional third argument to `divide`, which must be either a quoted variable or a variable with no assigned value. The third argument is assigned the quotient if `divide`'s value is `true`; otherwise, it remains unchanged. In this way "full evaluation" can be applied uniformly to all arguments of the function, yet auxiliary information can still be passed back through some of them.

Example 51
Use of quoted variables as parameters to library procedures

Use a as a programming variable.

```
> a := 3:
> p := x^2 - 1:
> q := x+1:
```

divide returns true if the second argument divides the first over the rationals. It assigns the third argument the value of the quotient. The third argument should be either an unassigned variable or a quoted one.

```
> divide(p, q, b);
                                    true
> b;
                                   x - 1
```

In this divide computation, full evaluation of the arguments means divide(x^2-1, x+1, 3) is being specified. Assignment of the quotient to "3" is impossible, since "3" isn't a variable.

```
> divide(p, q, a);
Error, wrong number (or type) of parameters in function divide;
```

Evaluating ´a´ yields the name
of the programming variable a.
Thus, this example calls `divide`
with the arguments (x^2-1,
x+1, a).

```
> divide(p, q, 'a');
                                    true
> a;
                                   x - 1
```

2.7   Concatenation — forming new names from old

In Maple, concatenation is the joining together of two or more symbols to produce a single symbol. The
way concatenation is specified is through the (.) operator. For example, if u, t, and k are symbols with no
assigned values, then the expression u.t evaluates to the symbol ut, and the expression u.t.k evaluates to
the symbol utk. If the result of concatenation is a name, then you can have a concatenation expression on
the *left-hand side* of an assignment.
    Concatenation generally follows the full evaluation rule *except* that the leftmost symbol is not evaluated.
Example 52 illustrates the concatenation of symbols with values.

Example 52
Concatenation of symbols into a name

Concatenation creates one name
by joining several mathematical
symbols together.

```
> a . b;
                                     ab
> a . b . c;
                                    abc
```

If the symbols have assigned
values, then full evaluation
applies for all but the first
symbol.

```
> a:=x;
                                   a := x
> b:=2;
                                   b := 2
> c:=y;
                                   c := y
> concatresult := a . b . c;
                         concatresult := a2y
```

If some symbols do not evaluate
to symbols, then an expression
called an "unevaluated
concatenation" is formed.

```
> c := y + 1;
                                 c := y + 1
> concatresult := a . b . c;
                      concatresult := a2.(y + 1)
> y:=5;
                                   y := 5
> concatresult;
                                    a26
```

If the concatenated result is a variable name, then one can use it in an assignment statement on the left-hand side.	`> a.b := a * b;` $$a2 := 2\ x$$

`` `` `` (the name with no characters in it) can be used when you need evaluation of all symbols in a concatenated expression.

```
> `` . a . b . c;
                    x26
```

A succinct way to create the expression a0 + a1*x + a2*x^2 + ... + a7*x^7. Quotation is required in this example to prevent the premature evaluation of a.i before i takes on specific values.

```
> sum( 'a.i*x^i', i = 0..7);
              3      3      4      5      6      7
  a0 + a1 x + 2 x  + a3 x + a4 x + a5 x + a6 x + a7 x
```

1.3 and 1 .   3 both refer to the floating-point number. In this case, the dot is not a concatenation operator, but rather it is a decimal point.

```
> 1.3 * 1 . 3;
              1.69
```

Using the "if statement" (discussed fully in Chapter 3), we can demonstrate the difference between the name `13` and the number 13.

```
> a:= ``.1.3;
              a := 13
> if a=13 then yes else no fi;
              no
> if a=`13` then yes else no fi;
              yes
```

2.8   Looking at parts of expressions — op, nops, coeff

Every Maple expression can be categorized as being of one or more *types*. For example, $x + y + z$ is a sum (referred to as `` `+` `` to avoid confusion with the sum command), while $2*x*y$ is of type `` `*` ``. There are distinct types for the other conventional mathematical objects Maple can represent, as described in Tables 16 and 17. Table 18 lists some more general data structures that also can represent mathematical objects.

Every Maple expression is kept in a data structure. If the data structure contains several pieces of information, the structure may have parts. At the top level of information, these parts are arranged in a linear order, i.e. there's a first part, a second part, etc. However, each part may be a complete data structure itself.

TABLE 16
Types of mathematical objects in Maple

Object type; Instances	Maple type name, whattype result	Examples
number		
integer	integer	5
rational	fraction	1/3
floating-point	float	4.7
algebraic expression		
variable	string	x, `A string`
indexed name	indexed	a[i]
sum	`+`	x+y, x-y
product	`*`	x*y, x/y
power	`^`	x^5, 1/x
function	function	sin(1), f(x,y)
composition of functions	function	sin@cos, f@@5
user-defined infix* operators	function	a &+ b
operator	procedure	(x,y)->sin(x)*cos(y), D
equation	`=`	a=b-3
range	`..`	1..10
series	series	result of series
matrix (array)	See Section 2.13, "Linear algebra in Maple."	

*See ?neutral to read more about user-defined infix operators.

TABLE 17
Types of Boolean objects in Maple

Object type; Instances	Maple type name, whattype result	Examples
Boolean;		
or	`or`	a or b
not	`not`	not b
and	`and`	a and b
inequalities;		
less than	`<`	a<b; b>a
less than or equal	`<=`	a<=b; b>=a
not equal	`<>`	a<>b
equality	`=`	a=b-3

For example, $x + y + z$ is represented in a data structure. The data structure can be thought of as having three parts, each part holding the information about a term of the sum — in this case the names (symbols) x, y, and z. The expression $2 * (x + y) * z$ also has three parts, but its second part consists of a sum data structure that has two parts itself.

Maple has several functions to allow you to get at the pieces and type of any expression.

op(i, *expression*) refers to the ith part of an expression. For example, `op(1, 2*x*y)` is 2.

nops(*expression*) is the number of parts (operands) in the expression. For example, `nops(2*x*y)` is 3, while `nops([4,3])` is 2.

whattype(*expression*) returns a description of an expression's type. `whattype(2*x*y)` is `` `*` ``.

type(*expression*, *typename*) returns `true` or `false` depending upon whether the expression is of type *typename*. In addition to the `whattype` typenames, there are types that combine the various categories. For example, `type(2*x*y, `*`);` is true. `type(2*x*y, polynom)` is also true, since $2xy$ is a polynomial as well as a product. In Section 2.9.2, we will see how other Maple data structures such as lists and sets have types and op-parts. In Sections 3.8 and 3.9 we will look at Maple's more complicated types and how you may define your own types.

• *ATTENTION*

> The names of types can be assigned too, as they are no different from any other symbol names. Thus, if you use `integer := 5;` to label an integer result for further reference, and subsequently enter `type(36, integer);` "full evaluation" turns this into `type(36, 5);`. This causes an error since "5" is not a predefined type. Assignment to "system-reserved" names such as `integer`, or `polynom` should be avoided, since even if you avoid using `type`, a built-in command might use it in its internal programming. Maple has many such names. You can use `mint`, discussed in Section 3.16.1, to discover if you are using any system-reserved names. If you are unsure about whether you have used type names as variables, you can quote them to be safe, as in `type(36, 'integer');`

Example 53
Types and parts of expressions via `whattype` and op

```
> s := x+2*y+sin(3*z^2);
```
$$s := x + 2\,y + \sin(3\,z^2)$$

whattype describes the type of its argument. The type of an unassigned variable is string.

```
> whattype(s);
```
$$+$$

```
> whattype(t);
```
$$string$$

In its simplest form, the type
function has value true if its first
argument is of the type
described by its second
argument.

```
> type(1/2, 'integer');
                                    false
> type(5, 'constant');
                                    true
> type(x, 'name');
                                    true
```

nops(expr) is the number of
operands in the expression.
There are three terms in the sum
s. Two of the terms have
subterms.

```
> nops(s);
                                    3
```

/

op(i, expr) is the ith operand
of the expression.

```
> op(1, s);
                                    x
```

Operand 0 for a function
invocation is the name of a
function.

```
> nops(op(3,s));
                                    1
> op(0, op(3, s));
                                    sin
```

sin(3*z^2) has only one
operand, namely, its argument
3*z^2.

```
> op(1, op(3, s));
                                     2
                                  3 z
```

An error results when we ask for
the second argument of
sin(3*z^2), since it doesn't
have one.

```
> op(2, op(3, s));
Error, improper op or subscript selector
```

Quotients are represented as products. Products always have the rational constant as the first operand (except when the constant is 1).

```
> q := x/(6*y);
                              q := 1/6 x/y
> type(q, `*`);
                                  true
> op(1,q);
                                  1/6
> op(2,q);
                                   x
> r := op(3,q);
                               r := 1/y
> type(r, `^`);
                                  true
> op(1,r);
                                   y
> op(2,r);
                                  -1
> `product without rational constant` := x*y;
          product without rational constant := x y
> op(1, `product without rational constant`);
                                   x
> op(2, `product without rational constant`);
                                   y
```

For a mathematical expression p in expanded form (one that has been processed by the **expand** or **collect** commands), **coeff**(p, x, n) is the coefficient of x^n in p. **lcoeff**(p, x) is the leading coefficient of the polynomial in x. **lcoeff**$(p, [x,y])$ returns the leading coefficient of the polynomial in x and y where x is considered the main variable. ([x, y] is an example of a Maple *list*. Lists are discussed further in Section 2.9.) An error is given if p is not a polynomial in the specified variables. **tcoeff** works similarly to **lcoeff** for finding the trailing coefficients of polynomials.

Example 54
Finding coefficients of variables with **coeff**, **lcoeff**, and **tcoeff**

The coefficient of x^4 for p is a polynomial in y and z. $\cos y$ does not appear in p, so its coefficient is 0. Asking for the coefficient of $(\cos y)^0$ (or z) in p is the same as asking for the terms of p that are free of $\cos y$ (or z).

```
> p := 3*x^4*y^4 + 2*x^4*y^3 + 2*x^3*y^3*z - y^2*z^5 + x*y^5;
           4  4       4  3       3  3     2  5       5
     p := 3 x  y  + 2 x  y  + 2 x  y  z - y  z  + x y
> coeff(p, x, 4);
                            4       3
                         3 y  + 2 y
> coeff(p, cos(y), 1);
                               0
> coeff(p, cos(y), 0);
           4  4       4  3       3  3     2  5       5
        3 x  y  + 2 x  y  + 2 x  y  z - y  z  + x y
> coeff(p, z, 0);
              4  4       4  3       5
           3 x  y  + 2 x  y  + x y
```

Produce the leading coefficient of p considered as a polynomial in x.	`> lcoeff(p, x);`	$$3 y^4 + 2 y^3$$

Find the leading coefficient of that result considered as a polynomial in x and y.	`> lcoeff(p, [x,y]);`	3

Of course, if y is considered the principal variable, we do not get the same results.	`> lcoeff(p, [y,x]);`	1

2.9 Expression sequences, sets, and lists

This section discusses three of the fundamental Maple data structures: expression sequences, sets, and lists. However, there are many other types of data structures as well, besides the mathematical ones described in Tables 16 and 17. They are listed in Table 18.

TABLE 18
General-purpose data structures in Maple

Object type	Maple type name	Examples
array	array	`a := array(1..10);`
table	table	`t := table();`
concatenation	`` `.` ``	`a.(m+1)`
expression sequence	exprseq	`1, b, c`
name	name	`x`
list	list	`[a,b,c]`
set	set	`{a,b,c}`
procedure definition	procedure	`proc(x,y) (x+y)^2 end`
unevaluated expression	uneval	`` ``x`` ``

2.9.1 Expression sequences

An *expression sequence* is simply an ordering of Maple objects separated by commas, such as `a,b,c`. Sequences can be assigned to variables as values, just as with any other Maple result.

Sequences arise often in Maple, in part because sets and lists (see Section 2.9.2) can be built out of sequences. Some Maple built-in functions return sequences for their results. For example, the `solve` function will sometimes return a solution sequence when there is more than one solution. When the `op` function is given one argument, it returns a sequence of *all* the operands of its argument. The arguments to a function are also a sequence. The special symbol NULL stands for the empty sequence, the sequence with nothing in it.

Full evaluation applies to the construction of sequences. Thus, if seq1 is assigned the sequence of unassigned variables a,b,c, then seq1,d is the sequence a,b,c,d, while seq1,NULL,seq1,NULL is the sequence a,b,c,a,b,c.

Since sequences occur so often in practice, there is a built-in seq procedure that efficiently creates a sequence parameterized by an index variable varying through a range of integers.

seq($f(i)$, i=*low..high*)	Construct a sequence varying over an integer range
seq($f(x)$, x=*expression*)	Construct a sequence varying over parts of an expression

The first form of the seq function generates an expression sequence formed by substituting values for i within $f(i)$ sequentially from the integer *low* to the integer *high*. The expression $f(i)$ can actually be any expression that contains i; it need not be only in the form of a function call. It may even be an expression that doesn't contain i, thereby replicating a fixed expression for the entire sequence.

It's often useful to create a sequence of expressions that varies over the operands, or subexpressions, of another expression. The second form of the seq function does exactly this. It constructs the sequence $f(op(1, expression))$, ..., $f(op(n, expression))$, where n is the number of operands in *expression*.

Example 55
Sequences and the seq function

seq works much like sum. Create a sequence of the numbers 12 down to 0.

```
> seq(12-i, i=0..12);
            12, 11, 10, 9, 8, 7, 6, 5, 4, 3, 2, 1, 0
```

An example of how to build up a finite sequence bit by bit.

```
> seq1 := sin(x), x=0;
                    seq1 := sin(x), x = 0
> seq1 := seq1, 5;
                    seq1 := sin(x), x = 0, 5
```

Sequences can be used as an argument to a function.

```
> t := taylor(seq1);
                          3        5
            t := x - 1/6 x  + O(x )
```

The series data structure has powers and coefficients as its operands.

```
> seq(op(i,t), i = 0..nops(t)) ;
            x, 1, 1, -1/6, 3, O(1), 5
```

Create a sequence of the 0th, 2nd, and 4th operands of the series above, using the built-in seq function.

```
> seq(op(2*i,t), i = 0..2);
            x, 1, 3
```

op(expr), where only one
argument is given, returns a
sequence of the operands of the
expression, equivalent to
op(1..nops(expr), expr).

```
> op( x + y + z );
```
$$x, y, z$$

op(i..j, expr) is a sequence of
the ith through j^{th} operands of
expr.

```
> op(1..4, t);
```
$$1, 1, -1/6, 3$$

2.9.2 Sets and lists

Sets, *lists*, *tables*, and *arrays* are other ways of collecting information in Maple. We discuss sets and lists in this section; tables and arrays are discussed in the next section.

A Maple *set* is written as a sequence surrounded by braces { }, the usual mathematical notation for a finite set.[4] Like most Maple objects, sets can be assigned as values to variables, and can appear as an argument to, or result of, a function. Maple has built-in operators for set unions, intersections, and set differences.

TABLE 19
Set operations in Maple

Mathematical operation	Maple operation
$a \cup b$	a union b
$a \cap b$	a intersect b
$a - b$ (set difference)	a minus b

The ordering of elements of sets is determined "by the system". As with the ordering of terms in a sum, the ordering of a set is consistent within a session, but varies from session to session and from machine to machine. Thus, in one session, a set may be printed {x,y,z}, while in another it may appear as {z,y,x}. If you wish a particular ordering of elements, consider using a list to hold your data instead.

---

[4] (* and *) are acceptable substitutes for "curly braces" in Maple.

Example 56
Set operations

```
> set1 := {a, b};
                              set1 := {b, a}
> set2 := {a, d, e};
                           set2 := {a, d, e}
> set3 := {b, e};
                              set3 := {b, e}
```

```
> u := set1 union set2;
                           u := {b, a, d, e}
> i := set2 intersect set3;
                              i := {e}
> d := set1 minus set3;
                              d := {a}
```

Like expression sequences, a Maple *list* preserves the order of data in the way the user originally specifies. Duplicate copies of the same value are preserved in lists, whereas they are eliminated in sets. On the other hand, the operations of union, intersection, and difference are not defined for lists in Maple.

A Maple list can be created by surrounding a sequence of Maple objects by left and right brackets, [and].[5]   To illustrate the distinction between sequences and lists in Maple, [a,b], [b,c] remains as a sequence of two lists, while (a,b) , (b,c) would automatically be merged into a single sequence a,b,b,c. One use of lists is to group objects together when they are to be considered as a single argument to a function.

The op function works for sets and lists as one might expect: the ith operand of a set or list will be the ith element of the object, with the ordering based on how you would see the set or list printed out. A Maple idiom for operations involving lists (or sets) is to create a sequence that is then turned into a list by wrapping brackets (or braces) around it. (See Example 57 below.)

An alternative notation for referring to a particular operand of a list, set, or sequence *L* is to use *L[integer]*, which is referred to as a *selection operation*. *L[range]* is an alternative notation for specifying a sequence of operands.

subsop(*component index* = *replacement*, *expression*)	Create an expression with a component replaced

The subsop command is useful for creating altered versions of a set, list, or other data structure with multiple operands. Example 57 illustrates its usage.

---

[5](| and |) are alternatives for [and] if you have difficulty entering brackets on your system.

- *ATTENTION*

> Use `subsop` to create a new list or set with a single changed component. (See Example 57.) If L has a list or set as a value, then a subsequent assignment of the form L[*integer*] := *expression do not* have the effect of altering a component of the structure. Rather, it causes the values of L to be lost completely.

Example 57
Merging lists, reversing a list, and changing a component of a list

```
> list1 := [1, 2, 3];
                          list1 := [1, 2, 3]
> list2 := [4, 5, 6];
                          list2 := [4, 5, 6]
```

Concatenate two lists together. We extract the contents of the sublists by using op(*list*), which produces the sequence of elements of *list*. Then we join the two sequences together and form a new list from that.

```
> list3 := [op(list1), op(list2)];
                list3 := [1, 2, 3, 4, 5, 6]
```

Reversing a list.

```
> n := nops(list3);
                              n := 6
> list4 := [ seq( op(n-i, list3), i = 0..(n-1)) ];
                list4 := [6, 5, 4, 3, 2, 1]
```

list3[i] is another way to refer to op(i,list3). Here is alternative notation to reverse a list.

```
> list5 := [seq( list3[n-i+1], i=1..n )];
                list5 := [6, 5, 4, 3, 2, 1]
```

The "sequence idiom" also works for sets.

```
> set6 := {list5[1..4]} intersect {op(list1)};
                          set6 := {3}
```

To change the 5th element of a list, use the subsop command.

```
> list6 := subsop(5=47, list5);
                list6 := [6, 5, 4, 3, 47, 1]
```

Assignment does not change the list; rather it creates a table data structure, losing the other information from the list. Maple tables are discussed in Section 2.10.1. To see the value of a table, you must use the Maple `print` command, discussed in Section 1.17.

```
> list6[5] := 48;

                        list6[5] := 48

> print(list6);

                        table([
                            5 = 48
                        ])
```

2.10   Tables and arrays — indexed collections of data

A Maple table or array is a data structure that allows you rapid and convenient access to any piece of a collection of data by a key or index. They are one of the few Maple data structures whose subcomponents can be altered in place. In contrast, changing a component of a list, say through the `subsop` command, generates a completely new list. In this section we discuss how to use tables, and some of the many kinds of applications where they are useful.

2.10.1   Tables

Maple has *tables* for collecting information in a tabular form. In the example below, we create a Maple table in which the table entries are numbers; each table entry is indexed by a color. Conceptually, we are thinking of an association between colors and numbers representing the wavelengths (in nanometers) of the colors, as in Table 20.

A particular table entry has an *index* which describes its position within the table. A particular value within the table is accessed using the *selection operation* using the notation *tablename*[*index*], where *tablename* is the name of the table and *index* is a name or an expression sequence. The *index* is sometimes referred to as the *subscript* in the selection operation.

In Maple, the "table creation function" is `table`. If a list of equations relating the indices and the table values is given as the argument, then the list is used as a specification of initial entries for the table. If no arguments are given to `table`, the table will be initially empty. Any Maple programming variable can be assigned a table as its value. Once assigned, that variable becomes the name of the table. The contents of an entire table can be displayed using `print`.

TABLE 20
Wavelength table

Color	Wavelength (nanometers)
red	610
blue	480
violet	450
green	520

Example 58
Creating and modifying tables

Assign the variable Wavelength a table object with some pre-defined entries. The indices of the table are colors and the components are numbers (wavelength in nanometers). We assume that the names of the colors have no assigned values, or else we would have to quote them.

```
> Wavelength := table( [red=610, blue=480, violet=450,
> green=520, black=0]);
                 Wavelength := table([
                                      blue = 480
                                      violet = 450
                                      red = 610
                                      black = 0
                                      green = 520
                         ])
```

Access a particular table entry.

```
> Wavelength[red];
                              610
```

Add another entry to the table by assigning the appropriate indexed component.

```
> Wavelength[orange] := 600;
                 Wavelength[orange] := 600
```

Alter the table by replacing a component.

```
> Wavelength[red] := 640;
                 Wavelength[red] := 640
```

Delete an element from the table by unassigning as one would an unsubscripted programming variable.

```
> Wavelength[black] := 'Wavelength[black]';
                 Wavelength[black] := Wavelength[black]
```

To display an entire table, use print.

```
> Wavelength;
                                    Wavelength
> print(Wavelength);
                        table([
                            blue = 480
                            violet = 450
                            red = 640
                            green = 520
                            orange = 600
                        ])
```

The `indices` function gives a sequence of indexes used in the table. The `entries` function gives a sequence of assigned values.

```
> indices(Wavelength);
            [blue], [violet], [red], [green], [orange]
> entries(Wavelength);
            [480], [450], [640], [520], [600]
```

Tables can have heterogeneous indices — numbers, variables, expressions, sequences, or any other valid Maple object. For example, assigning `Memo[Monday]` a value will create a table with the entry for Monday in it, and assign it to the variable `Memo`, if `Memo` doesn't already have a table for its value. If `Memo` does have a table for its value, then that table is updated.

Example 59
Another example of a Maple table

The table `Memo` has heterogeneous indices.

```
> Memo [Monday]  := `Nancy's birthday`:
> Memo [1, 2 ]   := `buckle my shoe`:
> Memo [e=mc^2]  := `Einstein's formula`:
> Memo [G]       := [`gravitational constant`, 9.8, 32]:
> print(Memo);
                table([
                    G = [gravitational constant, 9.8, 32]
                    (1, 2) = buckle my shoe
                    Monday = Nancy's birthday
                               2
                    (e = mc ) = Einstein's formula
                ])
```

If the variable `a` is assigned a table as its value, then the action `b := eval(a)` assigns `b` the *same table*. Later alteration of the table through either `a` or `b` alters the single version of the table. (In computer science terms, `a` and `b` point to the same object.) If `a` is assigned another table (or other Maple object such as a sum or equation), then `b` will still have a table as a value. This would not have been the case if we had just used `b := a;`, which makes *b* take on whatever the value of `a` is.

- *ATTENTION*

> Evaluation of programming variables that are assigned tables as values works differently than the usual "full evaluation". This exception to "full evaluation" is called *last name evaluation*. With "last name evaluation", the result of evaluation is the symbol that was the last in the chain of values just before the table was reached. Thus, if you assign the programming variable `Wavelength` a table as its value, and then perform the assignment `W12 := Wavelength`, the result of evaluating `W12` is (the symbol) `Wavelength`. However, the table will be the result of evaluating `eval(W12)`, and `print(W12)` will display the table (see Example 60).

Example 60
Tables as assigned values, last name evaluation

Create a table as before.

```
> Wavelength := table( ['red'=610,      'blue'=480,
>     'violet'=450, 'green'=520,
>                         'black'=0] );
                    Wavelength := table([
                                          blue = 480
                                          violet = 450
                                          red = 610
                                          black = 0
                                          green = 520
                                      ])
```

W12 and W13 have the same table as their value, but only the last name in the chain of assignments is displayed, not the table itself.

```
> W12 := Wavelength:    W13:=W12:
> W12;
                    Wavelength

> W13;
                    Wavelength
```

We use `eval(tablename)` to get the table.

```
> eval(W13);
                    table([
                        blue = 480
                        violet = 450
                        red = 610
                        black = 0
                        green = 520
                    ])
```

`print(tablename)` will print the table, although the result of the print function is `NULL`, not the table.

```
> print(W13);
                    table([
                        blue = 480
                        violet = 450
                        red = 610
                        black = 0
                        green = 520
                    ])
```

To create a distinct version of the table that will not be affected when the original is altered, use the built-in copy function.

Example 61
Copies of tables

| Create a table of the first few Fibonnaci numbers. The 6th entry is intentionally created incorrectly. | ```
> Fibonnaci_table[1] := 1: Fibonnaci_table[2] := 1:
> Fibonnaci_table[3] := 2: Fibonnaci_table[4] := 3:
> Fibonnaci_table[5] := 5: Fibonnaci_table[6] := 9:
``` |
| Assign b the same table, assign c a (separate) copy of it. copy is a built-in function for creating a separate copy of a table or array. | ```
> b := eval(Fibonnaci_table):
> c := copy(Fibonnaci_table):
``` |
| Alter an element of the table assigned to both b and `Fibonnaci_table`. | ```
> b[6] := 8;
                              b[6] := 8
``` |
| The table, as referenced by the other name, is changed. | ```
> Fibonnaci_table[6];
                              8
``` |
| But the separate copy is unaltered. | ```
> c[6];
                              9
``` |

2.10.2   Arrays

Arrays are a restriction of tables. Any array can have only *integer* subscripts. The range of permissible subscripts must be stated when the array is created via the **array** function, which is similar to **table**. In creating an array, you must specify the index range of integers for each dimension of the array. In addition, you can specify whether the indexing of the array is sparse, symmetric, anti-symmetric, or some other possibility of your own design.

Changing and copying arrays works as with tables. What distinguishes arrays from tables in general, other than the restriction to integer indices, is that each reference to an array element is checked to see that it is within the index range specified when the array was created.
```

Example 62
Maple arrays

Define a to be an array with 2
rows and 2 columns.

```
> a := array(1..2, 1..2);
                    a := array(1 .. 2, 1 .. 2, [])
```

Entries of a can be assigned
values. Unassigned elements of
an array retain their own name
as value.

```
> a[1,1] := 1: a[1,2] := 2: a[2,2] := 4:
```

As with other kinds of tables,
last name evaluation instead of
full evaluation holds. Use print
to view an array.

```
> a;
                                a
> print(a);
                        [   1     2 ]
                        [           ]
                        [ a[2, 1]  4 ]
```

Symbolic indices create symbolic
array references. When the index
expression is given a value, then
full evaluation produces an
actual array reference.

```
> i := 'i';
                              i := i
> b := a[i+1,i];
                        b := a[i + 1, i]
> c := 2*b;
                        c := 2 a[i + 1, i]
> i := 1;
                              i := 1
> b;
                           a[2, 1]
```

Here, a one-dimensional array
with a subscript range −1 to 3 is
created. In a manner similar to
tables, we define b[-1]'s value
to be t when we create the array.

```
> b := array( -1..3, [-1=t]);
                        b := array(-1 .. 3,, [
                            -1 = t
                             0 = b[0]
                             1 = b[1]
                             2 = b[2]
                             3 = b[3]
                        ])
> b[-1] + b[1];
                           t + b[1]
```

2.10.3   Sparse, symmetric and other special indexing schemes for Maple arrays and tables

`array( `*indexing function*`, ... )`	Create an array with an indexing function
`table( `*indexing function*`, ... )`	Create a table with an indexing function

Indexing functions allow you to specify that array or table elements should take on certain values by default. To employ one of the built-in indexing functions given in Table 21, give its name as the first component of the array or table definition. Example 63 gives an example of use of the sparse indexing function, which sets up a default value of zero for any array or table element not otherwise assigned a value. Other indexing functions, such as `antisymmetric` enforce certain relationships between array/table elements.

TABLE 21
Known indexing functions

Features	Maple Notation
Entries are assumed to be 0 unless otherwise noted.	sparse
All entries whose indices are equal under permutations are the same.	symmetric
All entries whose indices are even permutations of each other are the same. All entries whose indices are odd permutations of each other are additive reciprocals. Entries with repeated indices are 0.	antisymmetric
All entries on the diagonal are assumed to be 1. All other entries are assumed to be 0.	identity

It is possible to define your own indexing functions — for example, arranging that every element of a table or array be initialized to the empty set. See the *Maple V Language Reference Manual* Section 6.5.3 for further details.

Example 63
Indexing functions for arrays and tables, and the `indices` and `entries` functions

We use sparse indexing in a table to count the number of coincident objects in a list. We define a table of elements and their frequencies.

```
> List := [x,y,x,z,x,z,a,z,x,x,y,b,a,z];
        List := [x, y, x, z, x, z, a, z, x, x, y, b, a, z]
> F := table(sparse);
                    F := table(sparse,[])
> for i in List do
>   F[i] := F[i]+1
> od:
```

Since the table is sparsely indexed, if a letter did not appear in the list it will have a value of 0 in the frequency table F. How many a's, c's and x's were in List?	`> F[a],F[c],F[x];`	2, 0, 5

What are the indices that give non-zero values in table F and their corresponding entries?	`> indices(F);` `> entries(F);`	[a], [z], [y], [x], [b] [2], [4], [2], [5], [1]

2.10.4   What's in a table or array: `indices` and `entries`

As we have seen, tables and arrays do not need to have all their elements assigned a value before further use is made of them – a symbolic subscripted name is used in formulae involving unassigned array/table elements, or perhaps an indexing function will provide values for certain elements by default. The `indices` function describes which elements of an array or table have assigned values, while the `entries` function returns a sequence of the values assigned. Examples 58 and 63 illustrate these functions.

2.11   Converting from one structure to another

The data structures used to represent sums, products, sets, lists, and arrays are highly similar. Each can be thought of as a collection of objects written down in some order. It's possible to think of transforming a sum to a list by replacing the + signs with commas, and then putting brackets around the entire expression.

In some applications, it is desirable to do such conversions. We can do this by means of the `convert` function. The command

`convert( `*expression*`, `*result type*` )`	Convert from one data structure to another

will look in Maple's preprogrammed library of conversions to see if there is preprogrammed knowledge of how to convert *expression* (a list, set, sum, product, etc.) to the desired data structure, one of the Maple type names in Tables 16, 17, and 18. See the *Maple V Library Reference Manual* or `?convert` for a complete list of target types. If such knowledge is in the library, the conversion happens according to its specifications. If the knowledge isn't in the library, then you will see the message `Error, unable to convert`.

Example 64
Converting between data structures

Define a list named `testscores`.	`> testscores := [99, 88, 76, 37, 92, 96, 88, 92];` `           testscores := [99, 88, 76, 37, 92, 96, 88, 92]`

What is the average of these numbers? Convert the list to a sum; the automatic simplifier will add them up.

```
> convert(testscores, `+`);
                        668
> "/nops(testscores);
                       167/2
```

Use `evalf` for the "calculator" result.

```
> evalf(");
                    83.50000000
```

How many distinct scores are there?

```
> nops( convert(testscores, set) );
                         6
```

Convert a list into an array/vector.

```
> a := convert([x,y,z], array);
                 a := [ x, y, z ]
```

Convert it back into a list.

```
> alist := convert(a, list);
                  alist := [x, y, z]
```

Define a 3 × 3 identity matrix.

```
> A := array(1..3,1..3, identity);
        A := array(identity, 1 .. 3, 1 .. 3, [])
```

Convert it into a list of lists; the ith element of the list is row i of A.

```
> Alist := convert(A, list, list);
        Alist := [[1, 0, 0], [0, 1, 0], [0, 0, 1]]
```

2.12   The map function: performing the same operation on all elements of a data structure

map(f, *expression*)	Apply a function to each component of an expression
map(f, *expression*, arg_2, arg_3, ..., arg_n)	Apply a function with multiple arguments to each component of an expression

Maple built-in operations such as `evalf`, `normal`, or `mod` are defined for numbers or expressions but do the "obvious thing" when given a list or set as an argument — they return a list or set in which each element has had the built-in function applied to it. The function `map` is useful in situations where you wish to do the same for any structure and any function f. `map` creates an expression where the ith operand of the new structure is the result of applying f to the ith operand of the original data structure. The type of the created expression is the same as that of the original data structure, unless the automatic simplification subsequently changes it. Thus `map` works not only on lists and sets, but also on sums, products, equations, arrays, tables, etc.

Example 65
map applies a function to each component of a structure

Effect of mapping sin on a list,
set, and sum.

```
> L := [ 3/5, Pi/2, ln(3)];
                        L := [3/5, 1/2 Pi, ln(3)]
> map(sin, L);
                        [sin(3/5), 1, sin(ln(3))]
> s := {3*Pi/4, Pi, 11*Pi/4};
                        s := {Pi, 3/4 Pi, 11/4 Pi}
> s2 := map(sin, s);
                                              1/2
                        s2 := {0, 1/2 2    }
> expr := 1 + ln(3) + exp(4);
                        expr := 1 + ln(3) + exp(4)
> map(sin, expr);
                   sin(1) + sin(ln(3)) + sin(exp(4))
```

evalf maps automatically.

```
> evalf(L);
            [.6000000000, 1.570796327, 1.098612289]
> evalf(s2);
                        {0, .7071067810}
```

A little advanced Maple (see
Sections 1.18 and 3.5). Define
new functions f and g. $f(x)$ is x^2.

```
> f := x -> x^2;
                                       2
                        f := x -> x
```

$g(x)$ is the sequence x, x^2.

```
> g := proc(x) RETURN(x,x^2) end;
g := proc(x) RETURN(x,x^2) end
```

```
> L2 := [1,2,3,4];
                        L2 := [1, 2, 3, 4]
> s2 := {op(L2)};
                        s2 := {1, 2, 3, 4}
> A2 := convert(L2, array);
                        A2 := [ 1, 2, 3, 4 ]
```

Map the operations of f and g
onto the list $L2$, the set $s2$, and
the array $A2$.

```
> L3 := map(f, L2);
                        L3 := [1, 4, 9, 16]
> map(g, L2);
                   [1, 1, 2, 4, 3, 9, 4, 16]
> map(g, s2);
                        {1, 2, 3, 4, 9, 16}
> map(f, A2);
                        [ 1, 4, 9, 16 ]
```

Compute the geometric mean of
two lists, component-wise
through zip.

```
> zip( (x,y) -> sqrt(x*y), L2, L3 );
```
$$[1, \ 2 \ 2^{1/2} \ , \ 3 \ 3^{1/2} \ , \ 8]$$

The second form of map is useful when you wish to apply a function of several arguments to all the terms of an *expression* fixing all but the first argument at some predefined value. It creates a new data structure of the same type as *expression*, with $f(op(i, \ expression), \ arg_2, \ arg_3, \ \ldots, \ arg_n)$ for the ith operand.

zip(f, u, v)	Apply a function to pairs of components.

zip takes two lists (or one dimensional arrays) and produces a list (or array) whose i^{th} element is f(u[i],v[i]).

Example 66
zip and map with more than two arguments

Compute the indefinite integral
with respect to t for three
integrands in a list.

```
> map( int, [exp(t), ln(t), tan(t)], t );
```
$$[exp(t), \ t \ ln(t) \ - \ t, \ - \ ln(cos(t))]$$

Compute the union of the
solution sets of two equations.

```
> map(solve, {x^2-1=0, 5*x=4}, x);
```
$$\{-1, \ 1, \ 4/5\}$$

2.13 Linear algebra in Maple

2.13.1 Simple matrix and vector calculations with evalm

Maple uses the convention that a vector is an array of the form array(1..n), and a matrix is an array of the form array(1..m, 1..n), for positive integer values m and n. By assigning arrays of this form to variables, you can create expressions that many Maple linear algebra procedures will work on.

An easy way to have Maple perform matrix or vector arithmetic is to apply the evalm procedure on a matrix/vector expression, using operators from Table 22. Example 67 illustrates this simple form of matrix algebra.

Example 67
Simple matrix arithmetic with evalm

Define A to have a matrix value.

```
> A := array([ [1,2,3], [x,x,5], [1/2,4,-1]]);
```
$$A := \begin{bmatrix} 1 & 2 & 3 \\ x & x & 5 \\ 1/2 & 4 & -1 \end{bmatrix}$$

TABLE 22
Matrix operations with `evalm`

Operation	Maple notation	Example
matrix or vector addition	+	`evalm(A+B)`
matrix-matrix multiplication or matrix-vector multiplication	&*	`evalm(A &* v)`
matrix inverse	^(-1)	`evalm(A^(-1))`
matrix powering, integer n	^(n)	`evalm(A^3)`
scalar multiplication	*	`evalm( 2* A &* B - C &* (a*v))`
matrix identity	&*()	`evalm( A - lambda * &*() )`

Compute the matrix inverse of A.

```
> Ainv := evalm(A^(-1));
                    [        x + 20          28            - 10 + 3 x  ]
                    [ - 2 ----------     ----------    - 2 ----------  ]
                    [      23 x - 30     23 x - 30          23 x - 30  ]
                    [                                                  ]
                    [      2 x + 5          5             - 5 + 3 x    ]
            Ainv := [     ----------    - ----------    2 ----------   ]
                    [      23 x - 30      23 x - 30        23 x - 30   ]
                    [                                                  ]
                    [          x            6                x         ]
                    [   7 ----------    - ----------    - 2 ----------  ]
                    [      23 x - 30      23 x - 30          23 x - 30  ]
```

Verify that the inverse is correct.

```
> evalm( A &* Ainv );
                              [ 1   0   0 ]
                              [           ]
                              [ 0   1   0 ]
                              [           ]
                              [ 0   0   1 ]
```

Define v to be a vector and assign values to its first two components.

```
> v := array(1..3):  v[1] := 23: v[2]:=-y:

> print(v);
                              [ 23, - y, v[3] ]
```

Ordinary evaluation leaves matrix expressions alone.

```
> A &* v;
                                  A &* v
```

evalm uses matrix/vector arithmetic during evaluation of its expression.

```
> evalm(A &* v);
     [ 23 - 2 y + 3 v[3], 23 x - x y + 5 v[3], 23/2 - 4 y - v[3] ]
```

Note that &* can be a matrix-matrix or a matrix-vector multiplication, but it is not a vector dot product.

```
> evalm(v &* v);
Error, (in linalg[multiply]) vector dimensions incompatible
```

Vectors (arrays of one dimension) are treated the same as one-column matrices with the same number of rows. Scalar multiplication is ordinary *.

```
> w := array(1..3,1..1, [[v[1]], [v[2]], [v[3]]]);
```
$$
w := \begin{bmatrix} 23 \\ -y \\ v[3] \end{bmatrix}
$$

```
> evalm( v - 2*w);
```
$$
\begin{bmatrix} -23 \\ y \\ -v[3] \end{bmatrix}
$$

```
> evalm(A &* w  - A &* v);
```
$$
\begin{bmatrix} 0 \\ 0 \\ 0 \end{bmatrix}
$$

```
> evalm( w &* v);
Error, (in linalg[multiply])
matrix/vector dimensions incompatible
```

Matrix powering uses ^.

```
> A_result := evalm( 2 * A^2 );
```
$$
A\_result := \begin{bmatrix} 5 + 4x & 28 + 4x & 20 \\ 2x + 2x^2 + 5 & 4x + 2x^2 + 40 & 16x - 10 \\ 8x & -6 + 8x & 45 \end{bmatrix}
$$

● *ATTENTION*

> evalm is "add-on" evaluation, and does not replace ordinary Maple automatic simplification. This is why matrix multiplication uses &* instead of just *, since the automatic simplification for * described in Section 1.19 involving commutivity, 0, and 1 is never "turned off". The rules of automatic simplification also explain why evalm(A^0) returns a scalar 1 instead of the identity matrix — automatic simplification turns A^0 into 1 before evalm begins its calculations.

2.13.2   The `linalg` linear algebra package and `with`

The Maple library contains several *packages*, each package a collection of related functions. This section discusses one package, `linalg`. This package is a collection of functions that work with the arrays that represent vectors and matrices in Maple. The command `with(linalg)` defines shorter names that can be used to call upon the routines in the `linalg` package. For example, after issuing `with(linalg)`, we can call `hilbert` instead of `linalg[hilbert]`. Tables 23 and 24 describe the names of common vector and matrix operations. The chances are good that Maple's library has your favorite linear algebra operation even if it's not listed in the Tables. The *Maple V Library Reference Manual* and `?linalg` supply details about which functions are available and how they work.

TABLE 23
Matrix/vector operations

Operation	Maple notation
$u + v$ (vector)	`add(u,v)` or `evalm(u+v)`
$A + B$ (matrix)	`add(A,B)` or `evalm(A+B)`
$A \cdot B$	`multiply(A,B)` or `evalm(A &* B)`
Av	`multiply(A,v)` or `evalm(A &* v)`
$u.v$	`dotprod(u,v)`
$u \times v$	`crossprod(u,v)`
A^{-1}	`inverse(A)` or `evalm(1/A)`
$det\ A$	`det(A)`
$tr\ A$	`trace(A)`
$adj\ A$	`adjoint(A)`
A^T	`transpose(A)`
Permanent of A	`permanent(A)`

TABLE 24
Matrix/vector mathematical operations

Operation	Maple notation
Solve $Ax = b$	`linsolve(A,b)`
Perform Gaussian elimination on A	`gausselim(A)`
Reduced row echelon (Gauss-Jordan) form of A	`rref(A)` or `gaussjord(A)`
Eigenvalues of A	`eigenvals(A)`
Eigenvectors of A	`eigenvects(A)`
Characteristic matrix of A and x	`charmat(A,x)`
Characteristic polynomial of A	`charpoly(A,x)`
Minimal polynomial of A	`minpoly(A,x)`
Matrix exponential of A	`exponential(A)`
Jordan form of A	`jordan(A)`
Smith normal form of A	`smith(A,x)`
Hermite form of A	`hermite(A,x)`

Tables 23—28 contain only a partial list of the functions in `linalg`. The *Maple V Library Reference Manual* and `?linalg` supply more details about the linear algebra functions and how they work.

Example 68
The linear algebra package

Load in the linear algebra package. When loading with `with`, you see the names of all the functions you are reading in. To load the package silently, enter `with(linalg):` .

```
> with(linalg);
Warning: new definition for    norm
Warning: new definition for    trace
[BlockDiagonal, GramSchmidt, JordanBlock, add, addcol, addrow,
 adj, adjoint, angle, augment, backsub, band, basis, bezout,
 charmat, charpoly, col, coldim, colspace, colspan,
 companion, concat, cond, copyinto, crossprod, curl,
 definite, delcols, delrows, det, diag, diverge, dotprod,
 eigenvals, eigenvects, equal, exponential, extend,
 ffgausselim, fibonacci, frobenius, gausselim, gaussjord,
 genmatrix, grad, hadamard, hermite, hessian, hilbert,
 htranspose, ihermite, indexfunc, innerprod, intbasis,
 inverse, ismith, iszero, jacobian, jordan, kernel,
 laplacian, leastsqrs, linsolve, matrix, minor, minpoly,
 mulcol, mulrow, multiply, norm, nullspace, orthog,
 permanent, pivot, potential, randmatrix, range, rank, row,
 rowdim, rowspace, rowspan, rref, scalarmul, singularvals,
 smith, stack, submatrix, subvector, sumbasis, swapcol,
 swaprow, sylvester, toeplitz, trace, transpose, vandermonde,
 vecpotent, vectdim, vector]
```

| Define A to be a 3 x 3 matrix. | ```
> A := array( [[1,2,3], [3,-1,6], [4,8,12]] );
                     [ 1    2    3 ]
                     [             ]
              A := [ 3   -1    6 ]
                     [             ]
                     [ 4    8   12 ]
``` |

```
> u := array( [ -1, 3-s, -4 ] );
              u := [ -1, 3 - s, -4 ]
```

Define u to be a three-dimensional vector.

Solve A x = u. If there is more than one solution, linsolve uses the parameters t1, t2, t3, etc. to describe the solutions parametrically. In the answer produced by linsolve, any value of t2 would give a valid solution to the matrix equation.

```
> linsolve(A,u);
       [ 5 t2 + 5 - s, t2, - 7/3 t2 - 2 + 1/3 s ]
```

Compute the determinant of A.

```
> det(A);
                          0
```

Compute the characteristic polynomial of A.

```
> cpoly := charpoly(A,lambda);
                           3              2
         cpoly := lambda  - 12 lambda  - 67 lambda
```

Check out the Cayley-Hamilton theorem by evaluating the characteristic polynomial at lambda=A and verifying that it produces a matrix of all zeros.

```
> subs(lambda=A,cpoly);
                     3       2
                    A  - 12 A  - 67 A
> evalm(");
                       [ 0  0  0 ]
                       [         ]
                       [ 0  0  0 ]
                       [         ]
                       [ 0  0  0 ]
```

Compute the eigenvalues of A.

```
> eigenvals(A);
                        1/2          1/2
            0, 6 + 103    , 6 - 103
```

Factor the characteristic polynomial – it should have the eigenvalues of A as its roots.

```
> factor(cpoly);
                              2
            lambda (lambda  - 12 lambda - 67)
```
```

Compute the matrix exponential
of a simple anti-symmetric
matrix.

```
> a := matrix([[0, x], [-x, 0]]);
                  [  0    x ]
             a := [         ]
                  [ - x   0 ]
> exponential(a);
             [   cos(x)    sin(x) ]
             [                    ]
             [ - sin(x)   cos(x) ]
```

2.13.3   `linalg` procedures for vector spaces

Given a set or list of vectors V, the vector space *spanned* or *generated* by V is the (infinite) collection of vectors one gets by looking at all possible ways to add scalar multiples of vectors from V together. Sometimes the collection of vectors arises from looking at a matrix A and considering each row or column of V as a vector. In finite-dimensional linear algebra, V is called a *basis* if there is no other collection of vectors with fewer vectors than V that span the same vector space.

Table 25 describes some of the `linalg` procedures that calculate vector space bases or spanning sets.

TABLE 25
Operations that compute vector spanning sets or bases

Operation	Maple notation
Given a list or set of vectors V, find a subcollection of them that is a basis for the vector space spanned by V. Return them as a list (set).	`basis`(V)
Find a basis for the kernel (also known as the null space) of the matrix A – the vector space of vectors v such that Av is the zero vector.	`kernel`(A) `nullspace`(A)
Find a basis for the vector space generated by the rows (columns) of the matrix A considered as vectors.	`rowspace`(A) `colspace`(A)
Compute an orthogonal basis* for the vector space generated by the columns of the matrix A.	`GramSchmidt`(A)

*A collection of vectors is orthogonal if the dot product of any two distinct vectors taken from the collection is zero.

2.13.4   Differentiation applied to matrices and vectors

Scientific applications often involve vectors or matrices whose entries involve symbolic expressions. The `linalg` package supports many operations on such objects that involve differentiation or integration of their components. Table 26 describes the most common.

TABLE 26
Differential operations on matrices or vectors

Operation	Maple notation
Given f, a vector (list) of three expressions, and v a vector(list) of three variables, compute the curl of f with respect to v: $\nabla \times f = (\partial f_3/\partial v_2 - \partial f_2/\partial v_3, \partial f_1/\partial v_3 - \partial f_3/\partial v_1, \partial f_2/\partial v_1 - \partial f_1/\partial v_2)$.	curl(f,v)
Given f, a vector of expressions of variables in the list v, compute the gradient of f, $\nabla f = (\partial f_1/\partial v_1, \ldots, \partial f_n/\partial v_n)$.	grad(f,v) ,
Compute the divergence of f with respect to variable in v, div$f = \sum_{i=1}^{n} \partial f_i/\partial v_i$.	diverge(f,v)
Given a scalar expression s and a vector (list) of symbols v, compute the Hessian matrix – a matrix whose $(i,j)^{\text{th}}$ entry is $\frac{\partial^2 s}{\partial v_i \partial v_j}$.	hessian(s,v)
For a scalar expression s, and vector (list) of variables v, compute $\sum_i \frac{\partial^2 s}{\partial v_i^2}$.	laplacian(s,v)
Given a vector (list) f of expressions and one of variables v, compute a matrix whose $(i,j)^{\text{th}}$ entry is $\frac{\partial f_i}{\partial v_j}$.	jacobian(f,v)

Example 69
Differential matrix and vector operations in `linalg`

```
> with(linalg):
Warning: new definition for    norm
Warning: new definition for    trace
```

Define a scalar function f.

```
> f := 4*x*z - 5*y*x^3;
                                    3
                   f := 4 x z - 5 y x
```

We calculate the gradient ∇f with respect to x, y, z.

```
> gradf := grad(f, [x,y,z]);
                            2        3
         gradf := [ 4 z - 15 y x , - 5 x , 4 x ]
```

Define a vector field v over the variables x, y, z.

```
> v := vector(3, [ 4*x-3*x^3*y, 7*x*y*z^2+5*y^3,
>    4*x^2*y^2+2*x ]);
                3              2      3      2 2
      v := [ 4 x - 3 y x , 7 x y z  + 5 y , 4 x  y  + 2 x ]
```

Calculate the curl $\nabla \times v$,

```
> curlv := curl(v, [x,y,z]);
                 2                    2            2       3
    curlv := [ 8 y x  - 14 x y z, - 8 x y  - 2, 7 y z  + 3 x  ]
```

and the Jacobian.

```
> jacobian(v, [x,y,z]);
          [               2                 3                    ]
          [ 4 - 9 y x            - 3 x              0            ]
          [                                                      ]
          [                  2          2       2                ]
          [    7 y z      7 x z  + 15 y     14 x y z             ]
          [                                                      ]
          [      2                 2                             ]
          [ 8 x y  + 2        8 y x                0            ]
```

We calculate the Laplacian of f, and note why the notation for the Laplacian is $\nabla^2 f$. evalb evaluates the equation as a Boolean expression and returns true if it can determine that the two sides of the equation are equal.

```
> lapf := laplacian(f, [x,y,z]);
                 lapf := - 30 y x
> evalb(lapf = diverge(gradf, [x,y,z]));
                       true
```

We see the definition of the vector Laplacian is $\nabla(\text{div } v) - \nabla \times (\nabla \times v)$.

```
> vectlapv:= map(laplacian, v, [x,y,z]);
                                              2         2
        vectlapv := [ - 18 y x, 30 y + 14 y x, 8 y  + 8 x  ]
> graddivv := grad(divv, [x,y,z]);
                    graddivv := [ 0, 0, 0 ]
> curlcurlv := curl(curlv, [x,y,z]);
                 2              2        2      2
    curlcurlv := [ 7 z , - 14 y x - 9 x , - 8 y  - 8 x
                   + 14 x z ]
> evalm(vectlapv - (graddivv-curlcurlv));
                       2              2
      [ - 18 y x + 7 z , 30 y - 9 x , 14 x z ]
```

2.13.5   Structural operations on matrices and vectors

Many situations in linear algebra involve combining or operating on particular elements of a matrix or vector, often particular rows or columns. The linalg package contains many functions to help you perform such structural operations. Tables 27 and 28 describe the operations available. Example 70 illustrates these operations.

Example 70
Some matrix structure operations

```
> with(linalg):
Warning: new definition for   norm
Warning: new definition for   trace
```

```
> A := matrix( 3,3, [19,-50,88, 53,85,-49, 78,17,72] );
                            [ 19   -50    88 ]
                            [                ]
                     A := [ 53    85   -49 ]
                            [                ]
                            [ 78    17    72 ]
```

Of course, Maple has elementary matrix operations. We use some of these matrix operations to first subtract $A_{2,1}/A_{1,1}$ times the first row from the second row.

```
> B := addrow( A, 1, 2, -A[2,1]/A[1,1] );
                      [ 19       -50         88    ]
                      [                            ]
                      [          4265       5595   ]
                B := [  0      ----    -   ----   ]
                      [           19         19    ]
                      [                            ]
                      [ 78        17         72    ]
```

Then we swap rows 3 and 2.

```
> C := swaprow( B, 3, 2 );
                      [ 19       -50         88    ]
                      [                            ]
                      [ 78        17         72    ]
                C := [                            ]
                      [          4265       5595   ]
                      [  0      ----    -   ----   ]
                      [           19         19    ]
```

Finally, we multiply row 3 by 19.

```
> D := mulrow( C, 3, 19 );
                            [ 19    -50      88 ]
                            [                   ]
                     D := [ 78     17      72 ]
                            [                   ]
                            [  0   4265   -5595 ]
```

We can examine a submatrix.

```
> submatrix( A, 2..3, 1..3 );
                            [ 53   85   -49 ]
                            [              ]
                            [ 78   17    72 ]
```

We can put matrices together to form larger matrices.

```
> stack( B, row(C,1) );
                      [ 19       -50         88    ]
                      [                            ]
                      [          4265       5595   ]
                      [  0      ----    -   ----   ]
                      [           19         19    ]
                      [                            ]
                      [ 78        17         72    ]
                      [                            ]
                      [ 19       -50         88    ]
```

We are interested in solving `Ax=b` for several vectors.

```
> b1 := vector( 3, [3,5,-2] ):
> b2 := vector( 3, [4,-5,9] ):
> b3 := vector( 3, [21,4,7] ):
```

Using augment we create a new matrix and solve the equation for all three b's at once, by putting it into Gauss-Jordan form.

```
> augment( A, b1, b2, b3 );
        [ 19   -50    88    3    4   21 ]
        [                                ]
        [ 53    85   -49    5   -5    4 ]
        [                                ]
        [ 78    17    72   -2    9    7 ]
> gaussjord(");
     [                56399        42938       43729  ]
     [ 1   0   0      -----    -   -----       -----  ]
     [                 9855         9855        3285  ]
     [                                                ]
     [                20528        15761       15913  ]
     [ 0   1   0   -  -----        -----    -  -----  ]
     [                 3285         3285        1095  ]
     [                                                ]
     [                46832        36584       35782  ]
     [ 0   0   1   -  -----        -----    -  -----  ]
     [                 9855         9855        3285  ]
```

TABLE 27
Finding the dimensions of matrices and vectors

Operation	Maple notation
Number of rows of a matrix A.	`rowdim`(A)
Number of columns of A.	`coldim`(A)
Number of elements of a vector v.	`vectdim`(v)

TABLE 28
Structural operations on matrices and vectors

Operation	Maple notation
Create a matrix that is the same as A, except that the r_2^{th} row is the sum of row r_2 of A and m times row r_1.	$\texttt{addrow}(A,r_1,r_2,m)$
Similar to $\texttt{addrow}$ but for columns.	$\texttt{addcol}(A,c_1,c_2,m)$
Create a matrix that is the same as A except that the ith row (column) is multiplied by $expr$.	$\texttt{mulrow}(A,i,expr)$
Create a vector that is the ith row (column) of A.	$\texttt{row}(A,i), \texttt{col}(A,i)$
Create a matrix that has the columns of A and B (horizontal combination). $\texttt{augment}$ and $\texttt{concat}$ are synonymous.	$\texttt{augment}(A,B), \texttt{concat}(A,B)$
Paste the rows of A and B together (vertical combination).	$\texttt{stack}(A,B)$
Assign the elements of A to the $a_r \times a_c$ submatrix of B whose upper left corner is $B_{m,n}$.	$\texttt{copyinto}(A,B,m,n)$
Create a matrix that extends A with an extra m rows and an extra n columns. Initialize the extra elements to the scalar x if this optional argument is present.	$\texttt{extend}(A,B,m,n,x)$
Create a matrix consisting of rows $r_1..r_2$ and columns $c_1..c_2$ of A.	$\texttt{submatrix}(A, r_1..r_2, c_1..c_2)$
Create a vector consisting of elements $i..j$ of v.	$\texttt{subvector}(v, i..j)$
Create a matrix that is identical to A, except that rows (columns) i and j are swapped.	$\texttt{swaprow}(A,i,j), \texttt{swapcol}(A,i,j)$
Create a matrix that is the same as A, but has rows (columns) i through j deleted.	$\texttt{delrows}(A, i..j)$ $\texttt{delcols}(A, i..j)$
Create an n by n matrix that has the scalars d_1 through d_n on its diagonal.	$\texttt{diag}(d_1, d_2, ..., d_n)$
Create a matrix that has matrices B_1 through B_n as blocks on its diagonal.	$\texttt{diag}(B_1, B_2, ..., B_n)$
Create a coefficient matrix from a list (set) of equations and a list (set) of variables. If any third argument is present, the negative of the "right-hand side" vector of constant coefficients will be included in the matrix as its last column.	$\texttt{genmatrix}(equations, variables)$ $\texttt{genmatrix}(equations, variables, anything)$

2.13.6   Invoking specific functions from a package without `with`

You can avoid reading in all the functions that go with the command `with(linalg)` by using the "long form" `linalg[function] (...)`. For example, you can do `linalg[add] (u,v)` at any point in a Maple session without doing a prior `with`.

   `with(linalg,function)` defines only the particular procedure *function* instead of all the procedures in the package. Thus you can do `with(linalg,add);` and then just `add(u,v)`. This does not define `multiply`, `transpose`, etc. as `with(linalg);` would.

Example 71
More linear algebra, without `with`

Define I to be the 3 x 3 identity matrix.	```> I := array(1..3,1..3, identity);``` ```Error, Illegal use of an object as a name```

Oops, we forgot that I stands for the square root of −1. Let's try again.

```
> Id := array(1..3,1..3, identity);
        Id := array(identity, 1 .. 3, 1 .. 3, [])
```

The long-form of a `linalg` function works even with no prior `with` command.

```
> linalg[add] (Id,Id);
                    [ 2   0   0 ]
                    [           ]
                    [ 0   2   0 ]
                    [           ]
                    [ 0   0   2 ]
```

We can also use a form of `with` that only applies to the particular `linalg` procedure rather than everything in the package. `randmatrix(m,n)` creates a matrix with m rows and n columns with randomly chosen integer values.

```
> with(linalg,randmatrix);
                    [randmatrix]
> R := randmatrix(2,2);
                         [ -85   -55 ]
                    R := [           ]
                         [ -37   -35 ]
```

Defining a `linalg` function in this way still hasn't defined short (unqualified) names for other members of the `linalg` package.

```
> rref(R);
                       rref(R)
> linalg[rref] (R);
                        [ 1   0 ]
                        [       ]
                        [ 0   1 ]
```

- *ATTENTION*

> When you use `with` to make access to a package's functions convenient, you occasionally will override a previous definition of a function, either built-in or defined by you. For example, `trace` is by default defined to be the program-tracing function (see "`trace` and `printlevel`, program tracing tools" in Chapter 3). However, after doing `with(linalg)`, `trace` is redefined to mean the trace of a matrix. The `with` function prints out a warning message whenever it overrides a previous definition.

Besides `linalg`, there are packages for: beginning calculus students (`student`), statistics (`stats`), formal manipulation of power series (`powseries`), number theory (`numtheory`), special kinds of plotting (`plots`), linear programming (`simplex`), and orthogonal polynomials (`orthopoly`), among others. The `with` function and the "long-form" style can be used with any of the other packages in a manner similar to `linalg`. See the *Maple V Library Reference Manual* or `?packages` for further details.

2.14   `alias` for changing the names of built-in functions and mathematical symbols

`alias( ` *name* `= ` *expression* ` )`	Give an expression a nickname

It is natural to give short names to functions or expressions that you often use. One-letter (or one-symbol) names are standard practice, as in "let α stand for the expression $\sqrt{x^3 - 2}$". Maple uses the same process as `subs` to find where to substitute the alias. Assigning `alpha` instead of aliasing would not cause Maple to print out `alpha` in results.

Maple's `alias` command causes Maple to use the nickname when printing instead of the *expression* wherever it occurs. It also directs Maple to substitute expression wherever it sees the nickname used in interactive input.

Aliasing does not work inside the `alias` command itself, to allow convenient re-aliasing. Parameters and local variables in Maple procedures (discussed in Chapter 3) are not subject to aliases.

To see `alias` in action, see Example 45 in Section 2.3.7, and Section 6.5.

2.15   Saving the state of your Maple session

`save ` `` `filename.m` `` `;`	Save the state of a Maple session into a file
`save ` var_1, var_2, `...` var_n, `` `filename.m` `` `;`	Save the state of certain variables only
`read ` `` `filename.m` `` `;`	Load saved information into a Maple session

You can **save** the state of a Maple session, and continue with it later. The first form of the **save** statement will record the state of all variables and procedures in your file *filename*.m.

After a **save**, you can **quit** the Maple session. The file you've saved to will retain the state of your session. For example, `save ` `` `foo.m` `` `;` will save the state of the session on `foo.m`.

A **save** to any file with a name that ends in .m means that the results will be saved in *Maple internal format*. This format is the same as that used for the files of the Maple library.

- *ATTENTION*

> Don't forget the backquotes around "strange names" such as foo.m. If these surrounding back-quotes are missing, Maple will evaluate foo.m to foom since it sees two separate names that are separated by the concatenation operator (the period character).

What can you do with a file in .m format? You won't be able to see anything intelligible if you view it or print it out — Maple internal format is for the computer's convenience, not yours. The command **read** `` `filename.m` `` will set all variable and procedure definitions as they were when recorded in *filename*.m. You can use **save** to store your Maple work before you quit a Maple session, then use **read** in a subsequent session so that you can continue where you left off.

The second form of the **save** command will save only the state of the variables $var_1, var_2, \ldots var_n$. You can use this form of **save** if you wish to remember only certain portions of your work for later re-use. **read** `` `filename.m` `` will restore the variables to their recorded values.

Example 72
Retaining Maple results between sessions

In this example, % maple is the command to start up a Maple session.
% maple

```
   |\^/|       MAPLE V
._|\|   |/|_.  Copyright (c) 1981-1990 by the University of Waterloo.
 \  MAPLE  /   All rights reserved.  MAPLE is a registered trademark of
 <____ ____>   Waterloo Maple Software.
      |        Type ? for help.

> a := 47;

                               a := 47

> b := x^4;

                                 4
                            b := x

> #Save all values in the file FILEALL.m
> save `FILEALL.m`;
> #Save the value of a in the file FILESOME.m
> save a,`FILESOME.m`;
> quit;
bytes used=54276, alloc=65524, time=Float(50,-3)
```

In a later Maple session, reading in the file FILEALL.m recalls the values assigned in the previous session.

```
% maple

    |\^/|        MAPLE V
._|\|   |/|_.   Copyright (c) 1981-1990 by the University of Waterloo.
 \  MAPLE  /    All rights reserved.  MAPLE is a registered trademark of
 <____ ____>    Waterloo Maple Software.
      |         Type ? for help.

> read `FILEALL.m`;
> a*b;

                              4
                          47 x
```

FILESOME.m only had the value of a saved in it.

```
% maple

    |\^/|        MAPLE V
._|\|   |/|_.   Copyright (c) 1981-1990 by the University of Waterloo.
 \  MAPLE  /    All rights reserved.  MAPLE is a registered trademark of
 <____ ____>    Waterloo Maple Software.
      |         Type ? for help.

> read `FILESOME.m` ;
> a*b;

                          47 b

> quit;
```

Maple will automatically translate the file name you give in a **save** or **read** command (`` `FFT.m` ``, for example) to a filename appropriate to your computer system. On some systems this will be a different name (for example, FFT M for IBM CMS and FFT.M on VMS), while on others there will be no change (e.g., FFT.m on Unix. The Maple user can see what the translation is by invoking Maple's built-in **convert** function on the Maple filename.

convert(*filename*, hostfile)	Convert a Maple file name to the computer system's naming style

This command will produce the translation of the Maple *filename* appropriate for the computer system you're running on.

2.16   Recording results in files in human-readable format

save `` `filename` ``;	Save all results in a file using ordinary text
save var_1, var_2, ..., var_n, `` `filename` ``;	Save specified variables' values in a file, using ordinary text
read `` `filename` ``;	Read an ordinary text file of Maple commands

While internal format is designed to be quickly processed by the computer, it uses a code not easily read by people. Results written as text files of text are useful when you plan to use the results in another program that requires text input. You can also create or copy text files of Maple commands with a text editor or word processor, and have Maple perform the commands through reading the file.

When *filename* does not end in .m, the save command will save all current variables and their values using lprint format instead of Maple internal format. Giving a sequence of several variables before the filename (the second form of save at the beginning of this section) will save just the specified variables in the file.

You can read such files back into a Maple session[6] with read `` `filename` ``.

2.16.1   writeto and appendto: Maple session output to a file

writeto(*file name*)	Write Maple output into a file
appendto (*file name*) ;	Append Maple output to a file

The writeto command causes all printed output to be written into the file file name instead of the terminal. This includes the results of commands, of print and lprint, error messages, and so forth. The command writeto(terminal); returns the destination of output to your computer display.

The appendto command causes subsequent Maple output to be *appended* to the end of the specified file. Unlike writeto, appendto does not delete the existing contents of the file before writing starts.

2.16.2   Creating a transcript of a Maple session

Many computers have commands (external to Maple) that cause automatic recording of a session. On other computer systems (such as Macintosh and NeXT), the "worksheet" that the user interface presents for the format of the Maple session can be saved through a command selected from a menu. On other computer systems, it may be simply a matter of "cutting and pasting" from a window. Maple itself has no command for creating a transcript of an interactive session as you are creating it. While writeto does create a record of all input and output, it turns off output to the display while it is writing to the file.

Maple worksheets, on those systems that have them, are saved through a different mechanism than that of the save command. save stores variables and their values in a format that read can process as a sequence

---

[6]Unfortunately, lprint doesn't print out the backquotes for backquoted names. For example, lprint(`` `ab cd` ``); causes ab cd to be printed. So you're out of luck if you created a file in human-readable format with such names, and want to read the file in again.

of Maple commands. It is a "snapshot" of the variables with their values at the time of the **save** command. A saved worksheet stores all the actions of a session as word-processed text, Maple pretty-printed output, and graphs. Opening a saved worksheet does not set variables to the values mentioned within the worksheet.

2.17   Access to additional library procedures

readlib(*procedure name* **)**	Load a procedure from the Maple library into your Maple session

Many programming systems rely on a *library* to provide users with access to common procedures or data. The library is maintained as a collection of files that any user can access. The library is usually kept in a separate location from the files owned by users. The Maple library is not part of the initial contents of your Maple session, but pieces of it may be automatically read in during the session as the need arises.

Most of Maple's expertise in mathematical calculations is recorded in its library. To create Maple V's library, 5,779 files are used, a total of 14.4 million bytes that involve approximately 135,000 lines of programming and another 45,000 lines of **help** commentary. However, many of these files are omitted from the library when it is distributed. Those that are in the distribution include the library **.m** files (3,108 files, approximately 4.9 million bytes), and the files accessed by **help** and **?** (1,028 files, approximately 2.4 million bytes). You may find that your version of the Maple library differs in size or number of files from what we have described. Data compression, differences in file format, or custom-made libraries may contribute to the difference.

When you specify to Maple some common procedure such as **solve** or **sum**, the part of the library defining that procedure is loaded automatically into Maple (in the sense of the **read** statement) before the computation commences.   For example, when you issue the command

```
solve( x^2 - 1 = x , x );
```
Maple automatically reads the library file **solve.m** before the equation-solving computation starts.

During execution of **solve**, other files may be automatically read in. In this way specialized procedures for solving systems of linear equations, or for factoring polynomials, can be brought to bear on the problem as needed.

Maple provides *automatic loading* from the library as a user convenience for many commands such as **solve**. Automatic loading allows you to avoid **read** commands for most commonly-used library procedures. However, there are portions of the library that do not have automatic loading. The **?** (**help**) command and the *Maple V Library Reference Manual* will describe those library procedures that must be loaded in via **readlib** before they can be used.

If you wish to use a library procedure that does not read itself in automatically, **readlib**(*procedure name*) is the command to use. It is designed to be a convenient way of reading in library files that are not automatically loaded.

For example, **iroot** is a Maple library procedure that is not loaded automatically. (**iroot(x,n)** is a good integer approximation to $\sqrt[n]{x}$.) **readlib(iroot);** will load the appropriate file in — in this case, `` `.libname.`iroot.m`; ``.

Unlike **read**, **readlib** returns a value, the definition of the program that specifies how to compute the function being **readlib**ed. You can suppress the print-out of the program by using a colon instead of a semicolon after the **readlib** command.

Example 73
Reading in a library procedure with `readlib`

The library has a procedure for computing integer parts of nth roots. It is not loaded automatically.	`> iroot(103900,3);`	`iroot(103900, 3)`

The `iroot` library file can be read in via the `readlib` command. We use a colon to suppress the printing of the value returned by `readlib`.	`> readlib(iroot):`

Now we can use `iroot`.	`> iroot(103900,3);`	`47`

2.18   Other formats for output: `fortran`, `latex`, **and** `eqn`

Ordinarily, Maple displays results at the terminal using "prettyprinting". However, that isn't the only style of output we've seen. By using colons instead of semicolons we can suppress the printing of results of interactive commands. And as we found out in Chapter 1, we can print results at the terminal in "linear expression syntax" by using `lprint`, or into a file via `save`. In this section we discuss formats for output other than those of `print` or `lprint`. We also discuss commands beyond `save` that cause output to go into a file.

2.18.1   Writing expressions for Fortran and C programs

`fortran(`*expression*`)`	Print an expression in Fortran syntax
`fortran( `*expression*`, filename=`*file name* `)`	Print an expression in Fortran syntax into a file
`fortran( `*expression*`, optimized )`	Optimize an expression for evaluation in Fortran

The `fortran` command will print an expression using the syntax of the Fortran 77 programming language instead of that of Maple. If the expression is longer than one line, Maple will use Fortran continuation conventions (temporary variables `s1`, `s2`, etc. are used if the expression requires several Fortran statements to compute completely due to the 19-line limit for statements in Fortran). If *expression* is a list of Maple equations, then it will be translated as a sequence of Fortran assignment statements.

The second form of the `fortran` command will write the Fortran-style output on the specified file. The third form of the command takes a single expression and produces a statement sequence in Fortran format that reduces the number of arithmetic operations necessary to compute the expression. The improvement is accomplished through the use of common subexpressions and binary powering.[7] These two forms can be combined:

```
fortran( expr, filename=`prog.f`, optimized )
```

will write an optimized version of the expression in the file `prog.f`.

`readlib( C );`
`C( expression, optimized );`
`C( expression, filename=`prog.c`, optimized );`
	Print an expression using C syntax

The command for translating Maple expressions into the syntax of the programming language C is appropriately called `C`. Before using this command for the first time in a Maple session, you must use `readlib(C);` to load it in from the library. `C` requires `readlib` because there are many users wish to use the letter C for mathematical applications. If you want to use both `C` library function *and* the symbol C in a formula in a session, first rename the routine by `Ctrans := readlib(C):` and then use the symbol C and the procedure `Ctrans`.

Example 74
Fortran and C output displayed at the terminal

Since the Fortran assignment statement defining b contains a large expression, continuation lines are generated with a # in column 6.

```
> expr1 := arcsin(x) + 1/3*ln(x)^2 - ln(x)^3:
> fortran([a = expr1, b=expand(expr1^2)]);
      a = asin(x)+alog(x)**2/3-alog(x)**3
      b = asin(x)**2+2.0/3.0*asin(x)*alog(x)**2
     #-2*asin(x)*alog(x\)**3+alog(x)**4/9-2.0
     #/3.0*alog(x)**5+alog(x)**6
```

The global variable `precision` controls the precision of numbers and built-in functions. By default single precision is used, but double precision is also a possibility.

```
> precision := double;
                            precision := double
> fortran(expr1, optimized);
      t2 = dlog(x)
      t3 = t2**2
      t7 = dasin(x)+t3/3-t3*t2
> readlib(C):
> C([a= expr1], optimized);
      t2 = log(x);
      t3 = t2*t2;
      a = asin(x)+t3/3-t3*t2;
```

---

[7] `optimize`, a readlib-defined function, will display the sequence without translating it to Fortran format first.

2.18.2   Output formatted for LaTeX and eqn

latex and eqn are two readlib-defined functions that display expressions in a format suitable for entry into text processing programs. Here is an example:

Example 75
Generating a LaTeX version of an expression

Use the latex routine to produce	`> a := (4 + (x^2+1)/(sin(x)^2+x^2)) /`
LaTeX output for the expression	`>       (3/(sqrt(x)+1/(y+1)) +  binomial(3,n)):`
assigned to a.	`> latex(a, `latex.out`);`

The latex command in the example above produces the following result:

```
4+{\frac {x^{2}+1}{\sin(x)^{2}+x^{2}}}\left ({\frac {3}{\sqrt {x}+
\left (y+1\right )^{-1}}}+{3\choose n}\right )^{-1}
```

in the file latex.out.[8] Processing that file with LaTeX results in the output

$$4 + \frac{x^2+1}{\sin(x)^2+x^2}\left(\frac{3}{\sqrt{x}+(y+1)^{-1}}+\binom{3}{n}\right)^{-1}$$

The eqn command works very similarly to the latex command, only its output is to be used with the *troff* document processor.

Neither latex nor eqn try to solve the problem of deciding where to break an expression that requires more than one line to print. The user must manipulate the results of latex and eqn via a text editor to get the desired breaks.

---

[8]Commands that write on files such as latex use the same conventions as save to change "Maple-style names" into the file names as used by the rest of the computer system.

Chapter 3

The Maple Programming Language

We have already seen many examples of two key elements of Maple's programming language, the *expression statement* (a command or expression) and the *assignment statement*.

This chapter discusses the elements of more elaborate programming. It assumes that you are already familiar with another conventional programming language. Obviously, it is not intended to be an introduction to computer programming!

3.1 Repetition `while` you wait

`while` *condition*
`do`
    statement sequence
`od;`
	Repeat a sequence of statements while a condition is true

`while` causes repeated execution of its *statement sequence* until the *condition*, an expression that is either `true` or `false`, is no longer `true`. The while-do-od loop does not have to be spread out over several lines, but is usually more readable if it is laid out and indented appropriately over several lines.

Example 76
Use of `while` to compute integer g.c.d's

irem(m,n) is the integer
remainder of dividing the
integer m by the integer n.

```
> irem(5,3);
```
                              2

This programs Euclid's
algorithm for finding the greatest
common divisor of two integers.
There is a Maple built-in
function, `igcd`, which does this
for you, but here's how you
could get the computation by
giving the directions yourself.

```
> a := 20;
                                        a := 20
> b:=  12;
                                        b := 12
> while b<>0 do
>       d := irem(a,b);
>       a := b;
>       b := d;
> od;
                                        d := 8
                                        a := 12
                                        b := 8
                                        d := 4
                                        a := 8
                                        b := 4
                                        d := 0
                                        a := 4
                                        b := 0
> lprint(` integer gcd is`, a);
  integer gcd is    4
```

You can suppress the printing of
intermediate results in the
repetition by finishing your loop
with a colon.

```
> a := 35;
                                        a := 35
> b:=  15;
                                        b := 15
> while b<>0 do
>       d := irem(a,b);
>       a := b;
>       b := d;
> od:
> lprint(` integer gcd is`, a);
  integer gcd is    5
```

3.2   Repetition `for` **each one**

To repeat a sequence of commands for each of a sequence of values assigned to an index variable i, you can use the following:

`for` *i* `from` *start* `by` *change* `to` *finish*
`do`
    statement sequence
`od;`
	Repeat a sequence of statements for a sequence of values of *i*

The `for` statement sequence is repeated once for each distinct value of *i*. Initially, the index variable *i* is set to the *start* value. If *i* is less than or equal to *finish* (greater than or equal to if *change* is a negative value), then the *statement sequence* is performed. Then *i* is assigned the value *i* + *change*. The process of comparison and execution of the *statement sequence* is then performed repeatedly until the value of *i* goes beyond the value of *finish*. *i*'s final value will be its value that first exceeded *finish*.

● *ATTENTION*

> A common oversight is to use the variable i in a `for` statement and then to use it unquoted in a `sum` or `int`. This of course causes the value of i left over from the `for` to be used in the `sum` or `int`, instead of the intended (unassigned) symbol i. Remember to use quotes, or unassign your `for` loop index variables after you are finished with their values.

start, *change*, and *finish* must be expressions that evaluate to rational or floating-point numbers. If the `from` *start* or `by` *change* is omitted, a default value of 1 is used for the omitted value.

Example 77
for statement

Example of repetition with an automatically maintained index variable.

```
> for i from 2 by 2 to 6 do
>       lprint(`sum of `,j^i,` j = 1..n is:`);
>       print( expand(sum(j^i,j=1..n)) );
> od;
sum of    j**2    j = 1..n is:
                            3         2
                    1/3 n  + 1/2 n  + 1/6 n
sum of    j**4    j = 1..n is:
                        5         4         3
                1/5 n  + 1/2 n  + 1/3 n  - 1/30 n
sum of    j**6    j = 1..n is:
                            3         5         6         7
            1/42 n - 1/6 n  + 1/2 n  + 1/2 n  + 1/7 n
```

i's value after completion of the repetition is the next value it would take on if the repetition were to be continued.

```
> i;
                            8
```

It is often useful to look at each part of an expression in turn and to use these parts in some computation. One way of doing this would be to write

```
for i to nops(expression)
do
    statements using op(i, expression)
od
```

— all quite legal, but not that attractive. This programming idiom was showing up so often in Maple programs that another form of the for was introduced:

`for x in expression`
`do`
    `statements using x`
`od`
	Repeat for a sequence of values taken from an expression

Example 78
for-in statement

Here we compute the sum of
squares of even list members
using both kinds of for loops.

```
> aList := [1,2,3,4,5]:
> s := 0:
> for i to nops(aList) do
>       if irem(op(i,aList),2)=0 then
>            s := s + op(i,aList)^2
>       fi
> od:
> s;
                                    20
```

```
> s := 0:
> for n in aList do
>   if irem(n,2)=0 then s := s + n^2 fi
> od:
> s;
                                    20
```

The for and while can be combined. For example, the following loop will conclude with the value for i being the first prime number larger than 10^{10}.

```
for i from 10^10+1 by 2 while not isprime(i) do od;
```

3.3   Conditional execution with if-then-else-fi

`if condition then`
    `statement sequence 1`
`else`
    `statement sequence 2`
`fi;`
	Select a sequence of statements to perform

If you wish for *statement sequence 1* to be performed if *condition* is true, and *statement sequence 2* if it is not, then the `if` statement will do the trick.

Programming language etymologists will notice that Maple has a "do..od" and "if..fi" style of bracketing delimiters instead of "begin..end" or "if..endif". This is an idea borrowed from the programming language Algol 68.

The *condition* referred to above can be a combination of equalities, inequalities, variables, or functions that `evalb` would evaluate to true or false, such as `a+b<6` or `divide(poly1,poly2)` and `lcoeff(poly1)=1`.

Cases can be selected according to one of several conditions by

`if` *condition 1* `then`   Select one sequence to perform from several
    statement sequence 1
`elif` *condition 2* `then`
    statement sequence 2
 	
.
.
.
 	
`elif` *condition n* `then`
    statement sequence n
`else`
    default statement sequence
`fi;`

Example 79
Using `if` to print out results of primality testing

`isprime(n)` returns true if n is prime, false otherwise.

```
> isprime(37);
                              true
> isprime(6);
                              false
```

Inspect the numbers of the form $2^{(2n+1)} - 1$. Terminate the loop with a colon instead of a semicolon to display only the results of print.

```
> for i to 8 do
>     prime_candidate := 2^(2*i+1)-1;
>     if isprime(prime_candidate)
>     then print(prime_candidate, ` is prime`);
>     else print(prime_candidate, ` is not prime`)
>     fi;
> od:
                         7,  is prime
                        31,  is prime
                       127,  is prime
                       511,  is not prime
                      2047,  is not prime
                      8191,  is prime
                     32767,  is not prime
                    131071,  is prime
```

Example 80
if ... then ... elif ... else ... fi

```
> A := array( 1..4, 1..4 ):
```

Fill the upper triangle with the row indices, the lower triangle with the column indices, and the diagonal with 1's.

```
> for i to 4 do
>     for j to 4 do
>         if   i > j then A[i,j] := j
>         elif i < j then A[i,j] := i
>         else            A[i,j] := 1
>         fi
>     od;
> od:
> print(A);
                              [ 1   1   1   1 ]
                              [               ]
                              [ 1   1   2   2 ]
                              [               ]
                              [ 1   2   1   3 ]
                              [               ]
                              [ 1   2   3   1 ]
```

3.4   break and next: control within for-while loops

The value break, when encountered during a for-while repetition, causes immediate exit from the (innermost) for-while loop containing the break. An example of its use is in "search" situations, when you would like to stop looking through a series of possibilities as soon as you find what you are looking for.

When the value next is encountered during a for-while repetition, all statements are skipped from that point to the bracketing od of the (innermost) repetition loop. Execution then proceeds to the consideration of whether to perform another iteration.

Example 81
Leaving a for repetition early with break

Find the first non-prime in the sequence $2^i - 1$, $i = 3, 5, 7, ...$ When you hit a non-prime, break out of the repetition.

```
> for i from 3 by 2 do
>     if isprime(2^i-1)
>     then print(2^i-1, ` is prime`)
>     else break
>     fi;
> od;
                    7,   is prime
                   31,   is prime
                  127,   is prime
```

The value of i after the break is
its value when the break
occurred .

```
> print('i'=i, ` first non-prime in the sequence is`, 2^i-1);
            i = 9,  first non-prime in the sequence is, 511
```

3.5 Simple Maple procedures

proc(*sequence of formal parameters*) *an expression involving the parameter sequence* end	Define a Maple procedure (program)

We have already seen in Section 1.18 how to define "one-line" functions in Maple, such as `(x,y) -> x^2 + y^2;`. In this section, we discuss other ways of defining functions that can include a sequence of actions such as assignments, and `if`, `while`, or `for` statements.

Users can define their own functions by using `proc` expressions. All Maple procedures are functions, in that they are given values or names of objects as arguments, and return a value as a result. Procedures, like arrays, lists, and mathematical expressions, etc. are objects that are considered valid values for any programming variable to be assigned. Thus, a programming variable can be assigned a `proc` expression as its value. The name of the variable is the name of the function. Using a function defined through `proc` is the same as using any other function, with the "name of the function" being the name of the programming variable.

If you assign the variable `f` a `proc` expression, then you can call the procedure in the natural way — `f(3,5)`, for example. However, variables assigned `proc` expressions as values, like variables assigned arrays or tables, follow lastname evaluation rules, as discussed in Section 2.10.1. Thus evaluation of `f` will just produce `f`. To display `f`'s value, you can use the command `print(f);`. To access the `proc` object, you should use `eval(f);`.

One form of Maple function definition consists assigning a programming variable a `proc` expression. The *sequence of formal parameters* in a procedure definition must be a sequence of names. Within the `proc`-`end`, whatever values these symbols have outside the `proc` is ignored; they are used just as placeholders in describing what happens later when a computation is requested.[1]

One asks for a computation involving the function ("calling the function") by including in a command, or as part of a statement, the name of the function (the programming variable assigned the `proc` definition as a value), followed by a sequence of *arguments* enclosed by parentheses:

function_name(*expr1*, *expr2*, ...)

If the arguments involve programming variables as well as symbols or numbers, then "top-level evaluation" is applied to get values to be given to the function. Those values are then substituted into the *proc's expression*. The *proc's expression* is then evaluated and simplified. The result of evaluating and simplifying the expression is the value of the function. Example 82 illustrates the process of defining and using simple Maple functions.

---

[1] Procedure/function definition is one of the few exceptions to the general policy of "full evaluation".

Example 82
A simple Maple function defined using proc

Define a function/procedure f.
The sequence of formal
parameters is x, y. We could
have used the alternative
notation f := (x,y) -> x+y;
with the same effect.

```
> f := proc( x, y )
>   x+y;
> end;
f := proc(x,y) x+y end
```

Evaluate the function f with the
arguments 3, 5. That is, use 3 for
x and 5 for y when evaluating
x+y.

```
> f(3,5);
```
                                                                    8

Evaluate the function f at x=w,
y=z+1.

```
> f(w,z+1);
```
                                                            w + z + 1

An example of "last name
evaluation" – the value is not the
proc, but the last name in the
chain of values.

```
> g:= f;
```
                                                            g := f

```
> h:= g;
```
                                                            h := f

```
> h(3,5);
```
                                                                    8

```
> f;
```
                                                                    f

```
> g;
```
                                                                    f

```
> h;
```
                                                                    f

As with tables, op or print can
be used to view the proc object.

```
> op(f);
proc(x,y) x+y end
> print(h);
proc(x,y) x+y end
```

Conceptually, built-in commands such as diff or solve are like mathematical functions — given arguments, they return one value (expression) as a result. However, procedures can have effects beyond the values they return. For example, the Maple function divide has as an additional effect the assignment of a value to the programming variable given as its optional third argument. Another kind of side-effect would be a function that prints a "divide by zero" error message. Additionally, programming variables not included amongst the sequence of formal parameters may influence its results, such as the effect of Digits on evalf.

3.6   Maple procedures — multiple statements, local variables, RETURN

Sometimes it is not convenient to define the result of a procedure in terms of only one expression, as in the preceding section. The general form of a Maple `proc` definition is

```
proc( parameter sequence )                    Define a procedure with local variables and options
   local variable sequence;
   options option sequence;
   statement 1;
   statement 2;
      ⋮
   statement n
end;
```

The `local` and `options` sequences can be omitted if you do not need them. Among the `options` recognized by Maple is `option remember`.[2] We discuss it in Section 3.10 on page 133. We discuss local variables later in this section.

When Maple procedures involve several statements in their definition, the statements are performed one at a time, with the last value computed being the value of the procedure.

Example 83
A simple multiple statement procedure defined and used

A "toy" function for finding the maximum of three numbers. There is actually a Maple library function `max` which you could use instead.

```
> max3 := proc (a, b, c);
>     print(`Finding the maximum of`, a, b, c);
>     if a<b then
>         if b<c then c else b fi
>     elif a<c then c
>     else a
>     fi;
> end:
```

`max3(3,2,1)` executes the statements in the proc body with a having value 3, b having value 2, c having value 1.

```
> max3(3, 2, 1);
                Finding the maximum of, 3, 2, 1
                               3
> max3(1, 1, 9);
                Finding the maximum of, 1, 1, 9
                               9
```

Local variables are programming variables that are used only during execution of the `proc`, and then discarded. If a local variable has the same name as a variable known outside the procedure, then when the

---

[2]The other options are `builtin`, `operator`, `trace`, `arrow`, `angle` and `system`. See the *Maple V Library Reference Manual* or `?option` for further details on them.

name occurs in the statements defining the procedure, it refers to the local variable, not the exterior one(s).

Within a `proc`, you can retrieve the number of actual arguments specified through the special name `nargs`.[3]   The special name `args` has as its value the sequence of actual arguments. The ith argument in the sequence `args` is available by using the selection construct `args[i]`. This makes it possible to write procedures that are flexible in the number of arguments they take.

The special name `procname` has as its value the name of the procedure. It can be useful in invoking the procedure recursively or in creating expressions that involve the name of the procedure. Using `procname` instead of the name of the procedure will work even if the procedure is renamed, which might occur intentionally, inadvertently, or as a side effect of using Maple features such as `profile` (see Section 5.4 starting on page 189).

Example 84
Procedure using local variables

The example is a procedure to find the maximum of all the numbers given as arguments. Any number of arguments can be supplied.

```
> maxN := proc ()
>     local result, i;
```

In order to see if we can compute the maximum, we need to see if all the arguments are numeric — integer, rational, or float. We use the quoted procname expression to return the function call in case some of the arguments are non-numeric. Notice that we return the function call itself as a value, we place unevaluation quotes around it to prevent endless recursion. The behavior here is similar to what `int` or `sum` do when they can't return a closed-form solution.

```
>     if not( type([args], list(numeric)) )
>     then 'procname(args)';
>     elif nargs>0
>     then
>         result := args[1];
>         for i from 2 to nargs do
>             if args[i] > result then result := args[i]; fi;
>         od;
>         result;
>     fi;
> end:
> maxN(25/7, 525/149);
                              25/7
> maxN( 25/7, 525/149, 9/2);
                              9/2
```

maxN with no arguments returns the null sequence, since there was no "last value computed" in the function.

```
> maxN();
```

maxN returns itself if some of the arguments are non-numeric.

```
> maxN(25/7,z,525/149);
                                        525
                          maxN(25/7, z, ---)
                                        149
```

---

[3]`nargs` and `args` are *not variables,* because they cannot be assigned values in a procedure.

The local variable `result` has no assigned value outside of the procedure. In fact `maxN`'s local variable `result` is not even the same object as the `result` symbol.

```
> result;
```

$$result$$

RETURN(*value*)	Return a value from a procedure

When defining a multiple statement procedure, you may find that the required function value is already in hand after only some of the specified actions have been performed. It is possible to return the value at that point via RETURN. Example 85 illustrates the use of RETURN within procedures.

● *ATTENTION*

RETURN is different from the names `return` or `Return`. Spell it in all upper case letters.

Example 85
A procedure to compute Chebyshev polynomials using RETURN

`Chebyshev(n)` is a procedure to compute the Chebyshev polynomials of degrees 0 through n, returned in a table. We define the results for the first two Chebyshev polynomials explicitly. The rest are computed through a bootstrapping process.

```
> Chebyshev := proc (n)
>     local p,k;
>     p[0] := 1; p[1]:=x;
>     if n<=1 then RETURN( eval(p) ) fi;
>     for k from 2 to n do
>         p[k] := expand( 2*x*p[k-1] - p[k-2] )
>     od;
>     RETURN( eval(p) )
> end:
```

An example using the `Chebyshev` procedure.

```
> a := Chebyshev(5):
```

`a` now has as its value a table with six entries.

```
> seq(a[i], i = 0..5);
```

$$1, x, 2x^2 - 1, 4x^3 - 3x, 8x^4 - 8x^2 + 1, 16x^5 - 20x^3 + 5x$$

Display the value of the first six
Chebyshev polynomials at
$x = 1/4$.

```
> x := 1/4:
> seq(a[i], i = 0..5);
```

$$1, \ 1/4, \ -7/8, \ - \ \frac{11}{16}, \ \frac{17}{32}, \ \frac{61}{64}$$

Plot these polynomials on
$(-1, 1)$. First, though, unassign
x.

```
> x := 'x';
```

$$x := x$$

```
> plot({seq(a[i],i=0..5)},x=-1..1);
```

Local variables are evaluated using "one level evaluation". This means that when a local variable is evaluated, its value is the same as the last time it was assigned; its value is *not* re-evaluated to see if more simplifications can be done. This is normally what one wants when programming and from a practical point of view it leads to more efficient programs. If for some reason you want to fully (re-)evaluate a local variable, then you can use the eval function.

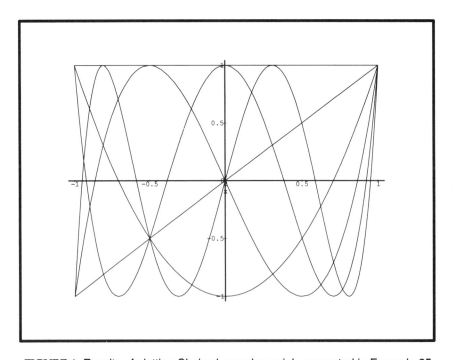

FIGURE 6. Results of plotting Chebyshev polynomials computed in Example 85

Example 86
Evaluation of local variables

<table>
<tr>
<td>

Evaluation of global names at top level uses full evaluation. The second evaluation of A uses the fact that w is now zero.

</td>
<td>

```
> A := sin(w):
> A;
                              sin(w)
> w := 0:
> A;
                                0
```

</td>
</tr>
<tr>
<td>

Evaluation of global names in a procedure is the same. It uses full evaluation too.

</td>
<td>

```
> Bfun := proc()
>     B := sin(x);
>     print(B);
>     x := 0;
>     print(B)
> end:
> Bfun();
                              sin(x)
                                0
```

</td>
</tr>
<tr>
<td>

Evaluation of local names is different: one-level evaluation is used. This means that the second evaluation of C will not see that y is zero. However, we can use eval to evaluate a local variable fully.

</td>
<td>

```
> Cfun := proc()
>     local C;
>     C := sin(y);
>     print(C);
>     y := 0;
>     print(C);
>     print(eval(C));
> end:
> Cfun();
                              sin(y)
                              sin(y)
                                0
```

</td>
</tr>
</table>

As we have mentioned, if you have a local variable x within a proc, the x within the proc has no connection with the usage of x outside the proc. A consequence of the way this is done in Maple is that local names can be used as mathematical symbols as well as programming variables, but although a global variable and local variable may have the same name and may be printed out the same, they are not the same!

● *ATTENTION*

Confusion between global variables and local variables with the same name can occur. The diagnostic program mint (discussed in Section 3.16.1) can help you to uncover such situations.

Suppose the procedure f calls or defines the procedure g within it. A statement within g cannot directly refer to any local variable of f (or vice versa, of course). A reference to a symbol x in g refers either to g's local variable x if it has one, or the "global" symbol x (the x that is defined at the interactive level, before or after any Maple procedures are used).

Example 87
The meaning of names when one procedure is defined within another

<table>
<tr>
<td>
The reference to y inside the procedure for g is to the global (top-level) variable y, not f's local variable of the same name.
</td>
<td>

```
> f := proc(x)
>       local y;
>       y := x + 1;
>       g := proc(a) a-y end;
>       g(x);
> end:
> f(5);
                              5 - y
> y := 47;
                              y := 47
> f(5);
                               -42
```

</td>
</tr>
<tr>
<td>
A demonstration that Maple is not dynamically scoped.
</td>
<td>

```
> h := proc(x)
>       local w;
>       w := x + 1;
>       i(x);
> end:
> i := proc(a) a-w end:
> h(5);
                              5 - w
```

</td>
</tr>
<tr>
<td>
The reference to w inside the procedure for i is to the global (top-level) variable w, not h's local variable of the same name.
</td>
<td>

```
> w := 47;
                              w := 47
> h(5);
                               -42
```

</td>
</tr>
</table>

3.7 Using ERROR — exiting several procedures at once

ERROR(`error message`)	Signal an error during execution

To indicate that something has gone wrong during execution of a procedure you have defined, you can use the ERROR function. (All the letters in ERROR must be capitalized, as with RETURN.) Executing an ERROR statement in a procedure prints a message of the form

    Error, (in *function_name*) *error message*.

As with all run-time errors, the computation is interrupted and aborted.

Example 88
Using the ERROR statement to signal a computational exception

Finding the maximum of some numbers as before, but this time an error occurs if no numbers are provided.

```
> maxN := proc ()
>     local result, i;
>     if nargs=0 then ERROR(`no arguments `) fi;
>     if not(type([args],list(numeric)))
>     then RETURN('procname(args)');
>     fi;
>     result := args[1];
>     for i from 2 to nargs do
>         if args[i] > result then result := args[i] fi;
>     od;
>     result;
> end:
> maxN( 25/7, 525/149, 9/2);
                                    9/2
> maxN( 25/7, z+1, 9/2);
                          maxN(25/7, z + 1, 9/2)
> maxN();
Error, (in maxN) no arguments
```

The `traperror` function allows you to avoid the interrupt/abort behavior due to errors in programs you write. As its name suggests, `traperror` is used to catch an error that is in the process of happening. It is typically used when you have reason to believe that a computation may cause an error to occur *and* you know what to do if one does. The variable `lasterror` records the nature of the last ERROR and may be used to control what the program does.

Example 89
`lasterror` and `traperror`

This computation normally causes execution to be interrupted.

```
> (1/3) mod 12;
Error, the modular inverse does not exist
```

The fact that this particular error occurred is recorded in `lasterror`.

```
> lasterror;
            the modular inverse does not exist
```

We can wrap the computations with `traperror`. These statements will either compute a modular inverse or generate a string containing an error message.

```
> u := traperror((1/4) mod 12);
                u := the modular inverse does not exist
> v := traperror((1/5) mod 12);
                                           v := 5
```

Example 90
Trapping an expected error

The following procedure illustrates how to trap an error in `normal` that occurs commonly. Since `normal` cannot always handle floating-point numbers, you may wish to use rationals.

```
> mynormal := proc(a)
>   local result;
>   result := traperror(normal(a));
>   if result=lasterror
>   then  if result=`floats not handled`
>         then result:=normal(convert(a,rational,exact))
>         else ERROR(result)
>         fi;
>   else result
>   fi
> end:
> mynormal((2.1+5.6*x)/(2.8+8.4*x));
                                      3 + 8 x
                                1/4  -------
                                      1 + 3 x
```

ERROR and `traperror` can also be used together to escape cleanly from a computation. For example, in computing an integer factorization, if a factor is discovered deep inside a subprogram one may want to return it immediately. By trapping errors selectively, one can escape from a computation without suppressing "legitimate" error messages. (Some other programming environments, including C-based and Lisp-based ones, allow this sort of non-local return.)

Example 91
Non-local returns using ERROR/`traperror`

Break out of a map with ERROR and `traperror`.

```
> List := [7,13,-8,2,-4,-9,5];
                List := [7, 13, -8, 2, -4, -9, 5]
```

We define a procedure to give an error if the input is negative. The error value is the negative number itself.

```
> err_neg := proc(x) if x<0 then ERROR(x) else x fi end:
```

This procedure will return the
first negative number in a list or
zero otherwise.

```
> first_neg := proc(a) local r;
>     r := traperror(map(err_neg,a));
>     if r=lasterror then r else 0 fi
> end:
> first_neg(List);
                                    -8
```

3.8   Checking types: writing safer programs

In Section 2.8 we saw that every object in Maple has a *type*, and that type(*expression*,*typename*) returns
true or false depending on whether *expression* is the proper type. Most of the mathematical types used by
Maple were listed in Table 16 on page 74: integer, rational, variable, sum, function, equation, and so forth.
Knowing something's type allows you to know what kind of operations you can perform on it. You can use
the Maple type function in procedures to check that the procedure's arguments permit the operations you
want to perform on them.

Typically the types of an argument to a procedure are checked with a statement of the form:

```
if not type ( argument, typename ) then
          ERROR(`wrong type argument to`, procname,`:`, argument,
          `should be of type`, typename );
fi;
```

This kind of type "guard" will permit execution of the rest of the procedure only if the argument is of the
correct type. *type checking* is good programming practice for most procedures.[4]

The following code fragment taken from Example 2 illustrates this process of simple type checking. We
check to see that the argument to the Fibonacci procedure is an integer. We also verify that the procedure is
invoked with only one argument. Furthermore, we guard against negative values being used as arguments
for this routine since that would cause it to loop forever.

```
if nargs<>1 or not type(n,integer) or n<0 then
          ERROR(`wrong number or type of parameters in F`)
fi;
```

3.9   Nested types and structured types

The types of Table 16 are known as *surface types*, since testing for such types needs only to look at type
information on the top level. However, there may be occasions when you may need to know more about an
argument than "surface type" testing permits. For example, you may wish to know not only if an argument

---

[4]The modest performance penalty incurred by type checking is usually worth the price because it reduces programming
or usage errors. Only when you are certain that a procedure will only be called with arguments of the correct type can
type checking be safely omitted. This might occur when the procedure is "internal" — to be called solely by other
procedures you have written where type checking is already in force.

is a polynomial, but that it is a univariate polynomial with integer coefficients in the variable x. Or, you may need to know if an argument is a power with a variable base and a positive integer exponent. Types in Maple come in two varieties that make it convenient to do this: *structured types* and *nested types*.

Structured types allow you to specify the form of the expression being type checked, and the type of each component of each form. Instead of a single *typename*, you use a Maple expression where each component of the expression is a type name or a structured type. Example 92 illustrates the use of a structured type.

Example 92
Definition and usage of structured types for type checking in procedures

Structured types make it easy to check for complicated expressions like the second parameter to sum.

```
> type( i=2..9, name=integer..integer );
                              true
```

A more complicated example might be checking for initial conditions in a differential equations solver.

```
> type( {f(1)=2,g(1)=3}, set(anyfunc(numeric)=numeric) );
                              true
```

Nested types typically cause inspection of all components of the type checked object in order to determine if the object is of the specified type. For example, the built-in nested type constant will cause type to inspect all the subcomponents of an expression to see if there are any parts that do not employ numbers or symbolic constants such as Pi at their lowest levels. Example 93 illustrates the use of nested types.

Example 93
Definition and usage of nested types for type checking in procedures

constant is a nested type. It knows that arithmetic on constants will also be a constant.

```
> 1-(Pi+f(2))^2;
                                   2
                       1 - (Pi + f(2))
> type(", constant);
                              true
```

polynom is also a nested type. With polynom you can specify the variables and the coefficient domain.

```
> p:=-x^2*y^6-4*x^2*z^4-2*x^2*y*z^4+3*x^5*y^2*z^2;
              2  6     2  4     2    4      5  2  2
     p := - x  y  - 4 x  z  - 2 x  y z  + 3 x  y  z
> type(p, polynom(integer,[x,y,z]));
                              true
```

For more information on built-in structured and nested types, see the *Maple V Language Reference Manual*, Chapter 5 "Type Testing", or ?type[structured] and ?type[surface]. A complete table of structured types can be found in the *Maple V Language Reference Manual* (Table 5.1, page 89).

3.10 Remembering function values

function (*argument sequence*) := *value*;	Remember a particular value for a function

You can remember particular values of a function by assignment. It will cause Maple to remember the "fact" that the value of *function* (*argument sequence*) is *value*. These remembered facts take priority over any definitions provided by a proc definition for the function.

The left-hand side of the assignment is not fully evaluated. Rather, the function name is evaluated and the arguments are evaluated but the function is not called. Example 94 illustrates a procedure that uses remembering by assignment.

Example 94
A function with "remembered" values

Example of use of remembering of a special value. Here we use the alternative syntax for procedure definition.

```
> f := x -> sin(x)/x;
                          sin(x)
              f := x -> ------
                            x
> f(0) := 1;
                       f(0) := 1
> f(infinity) := 0;
                    f(infinity) := 0
> [f(0), f(1), f(Pi/2), f(Pi), f(infinity)];
                              2
              [1, sin(1), ----, 0, 0]
                             Pi
```

As was mentioned in Section 3.6, option remember is one of the possibilities you can specify for a procedure when defining it. Whenever the value of a procedure with option remember is computed, the Maple system makes an entry in a table "remembering" the value of the procedure for the given arguments, just as if the value had been remembered through assignment. When the procedure is called again with the same arguments, the Maple system retrieves the result from the procedure's *remember table*. In this way, it is possible to avoid redundant computations. Example 95 defines and uses a procedure with option remember.

Example 95
A function with `option remember`

A way to compute the
Chebyshev polynomials of
Example 85 using option
remember instead of arrays:
Chebyshev2 takes a variable x
and an integer n. With option
remember, each Chebyshev
polynomial is computed once
and remembered internally.
Without remembering,
computing polynomials for large
values of n would involve
significant amounts of
redundant computation. We
need to call Chebyshev2 with the
same global variable XX_ each
time so that useful remembering
takes place.

```
> Chebyshev := proc(v,n)
>   local cpoly;
>   cpoly := Chebyshev2('XX_',n);
>   RETURN(subs('XX_'=v,cpoly));
> end:
>
> Chebyshev2 := proc(x,n)
>   option remember;
>   if n=0 then RETURN(1)
>   elif n=1 then RETURN(x)
>   else RETURN(normal(2*x*Chebyshev2(x,n-1)-Chebyshev2(x,n-2)))
>   fi;
> end:
```

We trace execution of
Chebyshev2 to verify that
redundant computations have
been eliminated. See Section
3.16.2 for more information on
program tracing.

```
> trace(Chebyshev2);
                              Chebyshev2
> Chebyshev(y,3);
--> enter Chebyshev2, args = XX_, 3
--> enter Chebyshev2, args = XX_, 2
--> enter Chebyshev2, args = XX_, 1
<-- exit Chebyshev2 = XX_
--> enter Chebyshev2, args = XX_, 0
<-- exit Chebyshev2 = 1
<-- exit Chebyshev2 = 2*XX_**2-1
<-- exit Chebyshev2 = 4*XX_**3-3*XX_
                              3
                          4 y  - 3 y
> Chebyshev(x,5);
--> enter Chebyshev2, args = XX_, 5
--> enter Chebyshev2, args = XX_, 4
<-- exit Chebyshev2 = 8*XX_**4-8*XX_**2+1
<-- exit Chebyshev2 = 16*XX_**5-20*XX_**3+5*XX_
                          5       3
                      16 x  - 20 x  + 5 x
```

● *ATTENTION*

> Remembering facts about procedures that make use of global variables can be difficult to do correctly. This is because the remembering mechanism only records the arguments and results of the procedure, not the value of global variables that were in effect during the computation. Example 96 illustrates how this can lead to erroneous results. Avoid using remembering with such procedures.

Example 96
A weakness of remembering — the value of relevant global variables is not recorded

We define a procedure for the Legendre polynomials. They are defined recursively as $L_n(x) = \frac{2n-1}{n}xL_{n-1} - \frac{n-1}{n}L_{n-2}$ with initial conditions $L_0(x) = 1$ and $L_1(x) = x$.

```
> L := proc(n,x)
>      option remember;
>      if n<2 then x^n
>      else normal( (2*n-1)/n*x * L(n-1,x) -
>                   (n-1)/n    * L(n-2,x) );
>      fi;
> end:
```

Now we try out our procedure.

```
> L(4, t);
```
$$35/8 \ t^4 \ - \ 15/4 \ t^2 \ + \ 3/8$$
```
> L(5, 4.33);
```
$$11284.20508$$

Let's obtain that last result to higher accuracy. The remember table doesn't know `Digits` has been changed, thus our previous result is echoed.

```
> Digits := 15;
```
$$Digits \ := \ 15$$
```
> L(5, 4.33);
```
$$11284.20508$$
```
> Digits := 10:
```

We can modify our procedure to only remember results when x is symbolic, not numeric. We explicitly assign symbolic results to the remember table.

```
> L := proc(n,x)
>      local result;
>      if n<2 then
>         result := x^n;
>      else
>         result := normal( (2*n-1)/n*x * L(n-1,x) -
>                           (n-1)/n    * L(n-2,x) );
>      fi;
>      if not type(x,numeric) then
>         L(n,x) := result;
>      fi;
>      RETURN(result);
> end:
```

Now we get the desired results.   `> L(4, y);`

$$35/8 \; y^4 - 15/4 \; y^2 + 3/8$$

`> L(5, 4.33);`

$$11284.20508$$

`> Digits := 15;`

$$Digits := 15$$

`> L(5, 4.33);`

$$11284.2050659470$$

The performance of extensive Maple computations can often be boosted by judicious use of `option remember`. See Section 5.5 for further information.

3.11   Functional operators

In Maple, functions can be assigned to variables in the same way numbers or expressions can be. We've already seen this with the assignment of user-defined procedures (`proc ... end` or "one-line functions"), but it also true for built-in functions. Maple can also perform limited amounts of algebra with functions — expressions consisting of built-in function names and procedure definitions can be summed, multiplied and exponentiated. These *functional expressions* can subsequently be evaluated at symbolic or numeric arguments. Functional expressions can be a succinct way of defining new functions from existing ones.

Some procedures are specifically designed to use functions as arguments, producing a function as a result. These are referred to as *functional operators*, and can also be a convenient way to define new functions from ones already defined. Three built-in functional operators are defined and explained in Table 29 and Example 97.

TABLE 29
Built-in functional operators

Operator	Effect of the operator
D (partial differentiation operator)	For a function f of one variable D(f) is the function that is the derivative of f. For a function of several variables g, D[1] (**g**) is the function that is the derivative with respect to the first argument, D[2] (**g**) is the derivative with respect to the second argument, and so forth.
@ (composition operator)	f @ g takes two functions (or operators) f and g as arguments and produces the function that is the composition of the first with the second. Thus (f@g)(x) gives the same result as f(g(x)).
@@ (repeated composition)	f @@ n for non-negative n, is equivalent to f composed with itself n times. Thus (f@@3)(x) is f(f(f(x))).

Example 97
Using functional expressions and the built-in functional operators D, @ and @@

D works on built-in function names, and on procedures.

```
> D(sin);
```
$$cos$$

The result of applying the functional operator D to a function is another function. Since the result is a function, it can be evaluated.

```
> f := x -> x^2;
```
$$f := x \rightarrow x^2$$

```
> df := D(f);
```
$$df := x \rightarrow 2\ x$$

```
> df(3);
```
$$6$$

If the derivative of a function is not known explicitly, then a symbolic formula is returned. Here we ask for the second derivative of a function g.

```
> (D@@2)(g);
```
$$D^{(2)}(g)$$

The symbolic form of the second derivative evaluated at 1.

```
> d2gAt1 := "(1);
```
$$d2gAt1 := D^{(2)}(g)(1)$$

Providing a value for g (a function) and re-evaluating provides the value of the second derivative at 1. Here we define $g(x)$ to be $\log(\tan x) + x^2$.

```
> g := log@tan + f;
```
$$g := log@tan + f$$

```
> d2gAt1;
```
$$-\ \frac{(1 + \tan(1)^2)\ 2}{\tan(1)^2} + 4 + 2\ \tan(1)^2$$

You can define your own functional operator by adding option operator to the options declaration part of the procedure definition. Example 98 illustrates the definition of a functional operator.

Example 98
Definition of a functional operator

Define a difference operator using the "functional template" method to create a new function. Note that the way to test to see if f is a Maple function is to use the type name procedure.

```
> Shift := proc(f)
>    local x;
>    option operator;
>    unapply( simplify(f(x+1) - f(x)), x);
> end:
```

By applying `Shift` to `sin` we get another function.

```
> SSin := Shift( sin );
                    SSin := x -> sin(x + 1) - sin(x)
```

Evaluate the function for numeric arguments.

```
> SSin(Pi);
                              - sin(1)
> SSin(1.0);
                           .0678264420
```

Evaluate the function for a symbolic argument.

```
> SSin(x^2);
                              2              2
                        sin(x  + 1) - sin(x )
```

Apply the Shift operator twice.

```
> Shift( (x) -> x^2);
                           x -> 2 x + 1
> Shift(");
                              2
```

A constant can also be given an argument, in which case it is interpreted as a constant function. This further explains the "zero-function trick" applied in to the asymptotic series of Example 34.

```
> "(3.5);
                              2
```

3.12   Packages in Maple

We have already discussed some of the packages available in Maple. Here we describe how one goes about writing a new package.

A package is simply a table in which the entries are procedures. We see that the "long form" of names for the things in the package are really just normal table references. For example, `simplex` is a package so `simplex[minimize]` is an entry from the table. Then `simplex[minimize](f,C)` calls the function entry in the table with arguments `f, C`.

A call to `with` reads a package in from the library and performs a number of assignments to make the functions available via their short names. When the command `with(simplex);` is given, an assignment of the form `minimize := simplex[minimize];` is executed, along with similar assignments for other package functions.

With this knowledge, it becomes fairly easy to write your own package. Example 99 illustrates this process. It reproduces a file of Maple definitions for quaternion arithmetic, complete with program comments and help for users. (Quaternions are 4-vectors that follow special rules for multiplication. Algebra texts such as [BM77] describe quaternions further.)

Example 99

A file for a simple Maple package

```
# Quaternion package.  Programmed by Greg Fee  April, 1992.
# The package consists of three operations: qparts, to extract the four
# coefficients from the data structure; qadd, to add quaternions, and
# qmul, to multiply them.  Quaternions are represented as an expression of
# the form quat(a0,a1,a2,a3).

quaternion[qparts] := proc(a,a0,a1,a2,a3)
        a0 := op(1,a); a1 := op(2,a);
        a2 := op(3,a); a3 := op(4,a);
        NULL
end:

quaternion[qadd] := proc(a,b)
local a0,a1,a2,a3,b0,b1,b2,b3;
        quaternion[qparts](a,a0,a1,a2,a3);
        quaternion[qparts](b,b0,b1,b2,b3);
        quat(a0+b0,a1+b1,a2+b2,a3+b3)
end:

quaternion[qmul] := proc(a,b)
local a0,a1,a2,a3,b0,b1,b2,b3;
        quaternion[qparts](a,a0,a1,a2,a3);
        quaternion[qparts](b,b0,b1,b2,b3);
        quat(a0*b0-a1*b1-a2*b2-a3*b3,
                  a0*b1+a1*b0+a2*b3-a3*b2,
                  a0*b2-a1*b3+a2*b0+a3*b1,
                  a0*b3+a1*b2-a2*b1+a3*b0)
end:


# The help information for the quaternion package.
help/text/quaternion := TEXT(

    ,
HELP FOR: Introduction to the quaternion package,

    ,
CALLING SEQUENCE:,
   <function>(args),
   quaternion[<function>](args),

    ,
SYNOPSIS:   ,
- To use a quaternion function, invoke the function,
   using the form quaternion[<function>].,

    ,
- The functions available are:,

   ,
                   qparts  qadd  qmult,
```

```
                ,
- For more information on a particular function see,
   quaternion[<function>].,

       ,
- As an example, to multiply two quaternions a and b, use,

       ,
       quaterion[mult](a,b);,

       ,
- Note that the quaternion functions expect as input,
   expressions of the form,
   quat( a0, a1, a2, a3).,

):


save quaternion.m;
quit
```

If you are in a new Maple session, then you need to read in the package first.

Example 100
Using your own package

```
                    > read `quaternion.m`;
```

Try out help for the package.

```
                    > ?quaternion
                    HELP FOR: Introduction to the quaternion package
                    CALLING SEQUENCE:
                        <function>(args)
                        quaternion[<function>](args)
                    SYNOPSIS:
                    - To use a quaternion function, invoke the function
                        using the form quaternion[<function>].
                    - The functions available are:
                                    qparts   qadd   qmult
                    - For more information on a particular function see
                        quaternion[<function>].
                    - As an example, to multiply two quaternions a and b, use
                            quaterion[mult](a,b);
                    - Note that the quaternion functions expect as input
                        expressions of the form
                        quat( a0, a1, a2, a3).
                    > a := quat(2,3,1,5): b := quat(9,4,7,6):
```

Multiply and add them together.

```
                    > quaternion[qmul](a,b);
                                          quat(-31, 6, 25, 74)
                    > quaternion[qadd](a,b);
                                          quat(11, 7, 8, 11)
```

3.13   Your Maple initialization file

If you have just one or two packages, you can read them in whenever you want to use them. If it turns out that you load them almost every time you use Maple, then you might like to have them read automatically when Maple starts up. You can do this with a Maple initialization file. In fact, you can have Maple execute any sequence of commands you like every time you start a new session.

The way you do this is by creating an initialization file containing commands to be read by Maple upon startup. For most computer systems, there are two types of Maple initialization files: a system-wide one and a personal one. When Maple is started, first it reads the system initialization file, if one exists, and then it reads your personal initialization file. After having performed all of the startup commands that have been placed in these files, Maple will then prompt you for input. The order in which the initialization files are read allows a group of users on a system to share common initialization commands and lets you override some of the common commands by resetting particular variables, redefining functions, etc.

What you put in this file depends on what you want to do. For example, to automatically change the default values for the global variables `Order`, `Digits`, and `prompt`,[5] put

    Order := 10:   Digits := 20:   prompt := `What now?`:

in your initialization file. It is a common practice to end all commands in a Maple initialization file with colons, as was done in the example above. This keeps the results of the commands from printing during initialization.

Maple expects the initialization file to have a particular name. This name depends on the kind of computer you have — see Table 30.[6]

The details of what a filename looks like will depend on the computer system and local custom. Maple automatically converts names such as `diff/GAMMA` into the name of a file within the Maple library directory in order to read in the definition of the derivative of the Γ function, for example. Most of the time the conversion does not amount to much — usually you can just type a file name as a string inside backquotes. However, on some computer systems you may have to use `convert`(*file name*, `hostfile`) (discussed in Section 2.15) to find out how to change a "Maple file name" (particularly one with a / in it such as `diff/GAMMA`) to a file name in the style of your computer system.

Getting back to packages, if the only package you had were `quaternion`, then you could have a Maple initialization file that contained the command to read in `quaternion.m` at the start of each session.

3.14   Creating `help` for your procedures

    `help/text/function` := TEXT(`text string 1`, `text string 2`, ..., `text string n`):

The TEXT data structure contains a sequence of character strings (left quoted names). When prettyprinted, each string is printed on a separate line without the left quotes. When you use the command ?*function* or

---

[5]The value of the variable `prompt`, if it has one, is used as the Maple prompt instead of the default.
[6]The system-specific documentation also explains how to suppress reading in the initialization file when you invoke Maple for those systems where it is possible.

TABLE 30
Name of Maple initialization file

Computer System	Type of initialization file	Name of initialization file
Unix	system	`src/init` in the Maple library directory
	personal	`.mapleinit` in your home directory
Macintosh	system	`MapleInit` in the Maple folder
	personal	`MapleInit` in your System folder
DOS	system	`MAPLE.INI` in the Maple LIB directory
	personal	`MAPLE.INI` in your current directory
DEC VMS	system	`MAPLE.INI` in the Maple LIB directory
	personal	`MAPLE.INI` in your current directory
IBM CMS	personal	`MAPLEINI FILE` on your `A` disk
All others	Consult "Getting started" and other system-specific documentation that came with your version of Maple.	

`help(`*function*`);`, the `help` procedure looks up the value of the variable `` `help/text/ ``*function*`` ` `` and if it has a value, prints it. For built-in library procedures, the appropriate variable has already been defined with a TEXT value. You can add help for your own procedures by defining the appropriate variables. Including the definition of the help variables within your package will ensure that ? help will be available for its procedures whenever it is read in.

It is probably easiest to create a help TEXT object by creating the "help description" as ordinary text using a word processor or text editor. When your description is ready, add the additional left quotes, commas, parentheses, and TEXT in the appropriate places.

The help text for Maple's built-in procedures follows a conventional format, using the following six categories:

HELP FOR: (Required.) The topic you are discussing.

CALLING SEQUENCE: (Required for procedures.) A description of how to invoke the procedure.

PARAMETERS: (Required for procedures and other Maple objects that use arguments or parameters.) A description of what each parameter must or can be in order for the procedure to work properly.

SYNOPSIS: (Required.) A brief description of how things work and what the purpose is of the item.

EXAMPLES: (Required for procedures and other objects for which there is a variety of valid arguments.) Give an example of how to invoke the procedure and what the result of invocation is. If the procedure has a variety of options, it will probably be worthwhile to give more than one example.

SEE ALSO: Mention related items for which help is available.

Example 99 contains a short help description.

3.15   Creating your own library

If you have a number of your own packages, you will probably wish to read them only when you intend to use them, since loading them all would be time-consuming and use memory that you could possibly wish to devote to your computation. You will want to start using your own library.

To use your own library, you have to tell Maple where to find it. The with command normally looks in the Maple library for packages.   This location within your file directory is specified by the global variable libname. If you have your own library then you can get with to look there too by assigning to the variable _liblist. (The first character in the name is an underscore.) You can either assign to this variable in every session or you can do it in your Maple initialization file.[7]

In general, the _liblist variable may refer to a list of libraries. If you share your computer with other people then you might want to have a private library and one common library. You should also include libname, a global variable whose value is the location of the built-in Maple library. Your initialization file could then contain something like:

```
mylibname     :=  `/u/mine/mymaplelib`:
sharedlibname :=  `/usr/local/maplelib`:
_liblist      := [mylibname, sharedlibname, libname]:
```

Now that you have told Maple where to find your library, you need to put your packages in it. The simplest way to do this is to move your ".m" files to the place specified by _liblist. On Unix, this would mean putting your ".m" files in the named directory. In general, the place for the ".m" file for the package mypack is given by the result of the Maple command

```
convert( . mylibname . /mypack.m, hostfile);
```

and the place for the source file is

```
convert( . mylibname . /src/mypack, hostfile);
```

Once you have done this, you can use with to load your packages and make the functions available by their short names.

---

[7]Even if you have permission to do so, it is not a good idea to add files directly to the built-in "Maple" library. That makes it harder to install new releases of Maple and you might overwrite something that really has to be there.

Example 101
With a package from your library

`libname` is a global variable whose value is the location of the built-in Maple library.	```\n> mylibname:= `/u/mapleleaves/mymaplelib`:\n> _liblist := [mylibname, libname]:\n```

Once `quaternion.m` has been added to `/u/mapleleaves/mymaplelib`, it becomes accessible via `with`.	```\n> with(quaternion);\n [qadd, qmul, qparts]\n>\n```

Three particular quaternions have special properties. Use the `alias` command to create nicknames for them, and use `qmul` to multiply them together. We can perform quaternion multiplication with just `qmul` because we have done a `with`.	```\n> alias(i= quat(0,1,0,0), j = quat(0,0,1,0), k=quat(0,0,0,1));\n I, i, j, k\n> qmul(i,j), qmul(j,k), qmul(k,i);\n k, i, j\n> qmul(i,i), qmul(j,j), qmul(k,k);\n quat(-1, 0, 0, 0), quat(-1, 0, 0, 0), quat(-1, 0, 0, 0)\n```

If a package is large, you may not want to load it all at once. To load the functions of a package one–by–one, as they are needed, you give `readlib` definitions for the individual entries.[8]

Example 102
Setting up a `readlib`-ed package

File `` `.mylib.`/src/quat ``	```\n> mylib:= `/u/mapleleaves/mymaplelib`:\n```

Rather than defining the procedures directly in this file, we only give `readlib` definitions.	```\n> quat := 'quaternion':\n> quat[qadd] :=\n> 'readlib(`quat/qadd`, ``.mylib.`/quat/qadd.m`)':\n> quat[qmult]  :=\n>    'readlib(`quat/qmult`,   ``.mylib.`/quat/qmult.m`)':\n> quat[qparts] :=\n> 'readlib(`quat/qparts`, ``.mylib.`/quat/qparts.m`)':\n```

---

[8]The cryptic-looking definitions of Example 102 work because of Maple's rules for full evaluation. When an entry of the table is used as a procedure name and invoked with arguments, full evaluation first causes the file to be read, and then invocation of the result of the `readlib` function — the procedure read in — to be called with the arguments. Since `readlib` uses `option remember`, no file reading will occur with subsequent invocations. Got it? Consider yourself on the way to becoming a Maple wizard!

```
> save quat, ``.mylib.`/quat.m`;
```

You also need to create one ".m" file for each function in the package.

Example 103
A member of a `readlib`-ed package

File
```.mylib.`/quat/src/qadd`

```
> mylib:=  `/u/mapleleaves/mymaplelib`:


> quaternion[qadd]  := proc(a,b)
> local a0,a1,a2,a3,b0,b1,b2,b3;
>   quaternion[qparts](a,a0,a1,a2,a3);
>   quaternion[qparts](b,b0,b1,b2,b3);
>   quat(a0+b0,a1+b1,a2+b2,a3+b3)
> end:
>
> save `quat/qadd`, ``.mylib.`/quat/qadd.m`;
```

With this definition and the value of _liblist established as in Example 101, all one needs to do is type
with(quat); and proceed as before.

3.16   Creating and debugging Maple programs

Every computer user has his/her own methods for creating an algorithm for solving a problem and verifying that it is correct. It isn't the point of this section to teach you how to devise efficient and correct programs. Once matters get to implementation in Maple, however, a typical scenario might be as follows:

(a) You write the solution in the Maple programming language.

(b) You enter your program into a file using a text editor or conventional means of text entry.

(c) You process the file using mint, the Maple program diagnostician. Using mint is a way to detect syntax errors and other common programming mistakes and some pitfalls before running the program.

(d) Using the read command, you load the code into a Maple session.

(e) With or without the program tracing tools printlevel and trace described below, you test the code. You may have to go back and rethink the algorithm and the code if you detect an error during testing.

(f) You achieve the desired result.

In this section, we discuss the tools available for Maple program development. We assume that by consulting the manuals for your computer system and by study of the previous section of this document (or by further reference to the *Maple V Language Reference Manual*), you have completed steps (a) and (b).
```

3.16.1   `mint`, the Maple diagnostician

`mint` will detect syntax errors in a file containing Maple code. It will also print a summary of usage of variables, built-in procedures, procedure parameters, and `ERROR` messages. This information can be used to detect programming that is legal, but probably incorrect, because of typographical errors, unintentional dual-purpose use of variables, etc.

   We illustrate the usage of `mint` below.

Example 104
File named `sample` to be processed by `mint`

Finding the maximum of some numbers as before, but this time an error occurs if no numbers are provided. There are numerous typographical errors, such as "results" instead of "result", and "kist" instead of "list". Assigning "list" (accidentally) will conflict with its use in other procedures in the context of type(y, list). "i" should be declared as a local variable. Also, the equation with LHS result was probably intended as an assignment.

```
> maxN := proc (x)
>    local list1, results;
>    list := [args];
>    if numargs=0 then ERROR(`no arguments`) fi;
>    result = kist1[1];
>    for i from 2 to nargs do
>       if list1[i] > result then result := list1[i] fi;
>    od;
>    result;
> end;
maxN := proc(x)
        local list1,results;
          list := [args];
          if numargs = 0 then ERROR(`no arguments`) fi;
          result = kist1[1];
          for i from 2 to nargs do
             if result < list1[i] then result := list1[i] fi
          od;
          result
        end
```


Example 105
Result of running file through `mint`

The command `mint<sample` processes the file sample with `mint` on a Unix system. The output of `mint` on `sample` is as follows:

```
     |\^/|
._|\|   |/|_.
 \  MINT  /   Maple V Diagnostic Program
  <____ ____>
      |
%
Procedure  maxN( x ) on lines 10 to 19
  These names were used as global names:   i, kist1, numargs, result
  These local variables were never used:   results
  These local variables were used but never assigned a value:   list1
```

```
These system defined names were assigned a value:    list
These names are special to Maple:
      list   defined as a type name, for the convert routines
These names were used as global names:    maxN
```

By default, mint describes "severe and serious errors" such as grammatically incorrect Maple programs. Assuming that the program you have given to mint does not have any grammatical errors, it will also describe the status of the symbols you use, such as the description of global and local variables in Example 105.

Certain kinds of mistakes turn out to be strictly speaking, grammatically correct – instances of when what you say is not what you really mean. The equation-as-a-statement in Example 104 is a case in point: It is perfectly legal to have any expression at all as part of a statement sequence. Someone used to programming in Fortran or C may find it helpful to have their attention called to the fact that Maple does not treat = as an assignment. Other typical mistakes include:

1. Mistaken expectations about the type of a variable – for example, expecting a list when in fact a set is returned.

2. Misspelling an occurrence of a local variable name (e.g. the use of kist1 instead of list1 in Example 105). While in many conventional programming languages this would be detected as a mistake because all variables must be declared, in Maple the misspelled variable is treated as a global symbol.

3. As a general practice, symbols used as programming variables within a program should be declared to be local. Symbols used as mathematical variables may be global, but they are typically passed into a procedure as one or as part of an procedure argument, rather than created within the procedure. Forgetting to declare a programming variable to be local, while not strictly speaking an error, often backfires because the global variable is changed in a command or during execution of another program.

As an option, mint can produce a more detailed analysis of your program file which will indicate the possible presence of such errors. Example 106 shows mint analyzing the file sample in this way for a Unix system. The -i 4 part of the mint command is the way on Unix to specify that you want the complete report (information level 4).[9]

Example 106
mint giving a full report on variable usage as well as errors and warning

```
%mint -i 4 sample
    |\^/|
._|\|   |/|_.
 \  MINT   /  Maple V Diagnostic Program
 <____ ____>
     |
Procedure  maxN( x ) on lines 10 to 19
   These names were used as global names:    i, kist1, numargs, result
```

---

[9]The directions for using mint at your computer installation may differ from those given above. See your local system documentation and consultants for details.

```
These local variables were never used:   results
These local variables were used but never assigned a value:   list1
These system defined names were assigned a value:   list
An equation is used as a statement on line 14
Parameter usage:
      Local variable usage:
    list1  was used as a list or table
      Global variables used in this procedure:
    i      was used as a value, assigned a value
    kist1  was used as a list or table
    numargs was used as a value
    result was used as a value, assigned a value
System names used in this procedure:
    ERROR  was called as a fcn
    args   was used as a value
    list   was assigned a list
    nargs  was used as a value
 Error msgs generated:
    no arguments
 Maple uses these names:
    ERROR  defined throughout Maple, for evalhf
    args   defined throughout Maple
    list   defined as a type name, for the convert routines
    nargs  defined throughout Maple
 These functions were called:   ERROR
Global variables used in this file:
    maxN   was assigned a procedure
```

3.16.2   `trace` and `printlevel`, program tracing tools

`trace` and `printlevel` are designed to provide step-by-step details on the execution of a Maple program, as it is executing. They can be used to locate exactly where execution "diverges" from your notion of correctness.

`printlevel` is the simpler of the two tools. It is a global variable (default value 1) that controls the amount of information printed out during execution. This information includes assignments, expressions, and procedure entries and exits performed during program execution. `printlevel` must be assigned an integer value. At its default value, just the results of statements typed into Maple, and statements nested one level inside `for/while` loops and `if` statements, are displayed. If it is assigned zero then the results of statements nested one or more levels are not displayed, and only direct statements typed into the session are displayed. If it is assigned a negative value then no results (not even those of simple commands) are displayed, except the output of `print` commands. The larger the value of `printlevel`, the more information is printed out on the results of function evaluations and other Maple statements nested within `for/while` statements, `if` statements, and procedures.[10]

---

[10]The information includes procedure entries (showing the arguments passed), procedure exits (showing the value returned), and the results of executing statements nested up to the specified level. The nesting of statements is counted in the sense of nestings within `for/while` statements and `if` statements as well as nestings within procedures. See the *Maple V Library Reference Manual* or `?printlevel` for a fuller explanation.

trace(*procedure name*) provides a way of looking only at the results of particular procedures. A traced procedure will behave as if printlevel was set sufficiently high as to print all the statements and entry and exit points of that procedure.

To turn off tracing of a procedure, use untrace(*procedure name*).

Example 107
Programming aids: printlevel

Example of printlevel set high enough to print out all intermediate results of a computation.

```
> printlevel := 11:
```

sum calls the built-in procedure `sum/indef`, which in turn calls other functions.

```
> sum(i^2, i=1..n);
--> enter sum, args = i**2, i = 1 .. n
--> enter type/RootOf, args = 1 .. n
<-- exit type/RootOf = false
                                xx := i
                                 a := 1
                                 b := n + 1
                                dab := n
                                dxx := 2
--> enter sum/indef, args = i**2, i
                                       2
                              tmp := _S
sum/indef: indefinite summation
sum/indef: indefinite summation finished
<-- exit sum/indef = 1/3*i**3-1/2*i**2+1/6*i
                     3        2
         ans := 1/3 i  - 1/2 i  + 1/6 i
                        3          2
         ans := 1/3 (n + 1)  - 1/2 (n + 1)  + 1/6 n + 1/6
                    3          2
         1/3 (n + 1)  - 1/2 (n + 1)  + 1/6 n + 1/6
<-- exit sum = 1/3*(n+1)**3-1/2*(n+1)**2+1/6*n+1/6
                    3          2
         1/3 (n + 1)  - 1/2 (n + 1)  + 1/6 n + 1/6
```

Reducing `printlevel` prints out information to a lesser depth of procedure calls.

```
> printlevel := 6:
> sum(i^3, i=1..n);
--> enter sum, args = i**3, i = 1 .. n
                          xx := i
                           a := 1
                          b := n + 1
                         dab := n
                         dxx := 3
sum/indef: indefinite summation
sum/indef: indefinite summation finished
                          4       3       2
              ans := 1/4 i  - 1/2 i  + 1/4 i
                 4          3          2
     1/4 (n + 1)  - 1/2 (n + 1)  + 1/4 (n + 1)
<-- exit sum = 1/4*(n+1)**4-1/2*(n+1)**3+1/4*(n+1)**2
                 4          3          2
     1/4 (n + 1)  - 1/2 (n + 1)  + 1/4 (n + 1)
```

Reducing `printlevel` back to its default value of 1 returns things to normal: just the results of the computation are printed.

```
> printlevel := 1;
                         printlevel := 1
> sum(i^4, i=1..n);
              5          4          3
   1/5 (n + 1)  - 1/2 (n + 1)  + 1/3 (n + 1)  - 1/30 n - 1/30
```

Example 108
Debugging with `trace`

Define a procedure to evaluate a specific function, in this case a polynomial at several values in a set. Return the results as a set.

```
> EvalPolyAt := proc(S)
>     local t, p, x, answer;
>     p := x^4 - 3*x^3 - 1;
>     answer := {};
>     for t in S do
>       x := t;
>       answer := answer union {p};
>     od;
>     RETURN(answer);
>     end:
> EvalPolyAt({2975, 5556, -1187});
                      4     3
                   {x  - 3 x  - 1}
```

It doesn't work. Let's investigate using `trace`. Note that although x is assigned a value it seems to have no effect on the value of p. We realize it's because there is only one-level evaluation of expressions within procedures.

```
> trace(EvalPolyAt);
                              EvalPolyAt
> EvalPolyAt({2, 3, 4});
--> enter EvalPolyAt, args = {2, 3, 4}
                            4       3
                   p := x  - 3 x  - 1
                        answer := {}
                          x := 2
                            4       3
                 answer := {x  - 3 x  - 1}
                          x := 3
                            4       3
                 answer := {x  - 3 x  - 1}
                          x := 4
                            4       3
                 answer := {x  - 3 x  - 1}
<-- exit EvalPolyAt = {x**4-3*x**3-1}
                         4       3
                       {x  - 3 x  - 1}
```

We can illustrate. p just evaluates to the polynomial and the x's in the polynomial are not evaluated.

```
> p := x^4 - 3*x^3 - 1;
                              4       3
                     p := x  - 3 x  - 1
> x := 4;
                           x := 4
> eval(p, 1);
                        4       3
                       x  - 3 x  - 1
```

However, in top-level Maple, full evaluation will allow the assignment to take effect.

```
> p;
                              63
```

Now we modify our procedure to use subs. We also note that our procedure returns a set of answers, which does not preserve the correspondence between input and output; we'll use a list instead. When we redefine `EvalPolyAt`, Maple automatically turns off the tracing of that procedure.

```
> EvalPolyAt := proc(S)
>     local t, p, x, answer;
>     p := x^4 - 3*x^3 - 1;
>     answer := NULL;
>     for t in S do
>        answer := answer, subs(x=t,p);
>     od;
>     RETURN([answer]);
>     end:
```

We see now that everything works fine.

```
> EvalPolyAt([2, 3, 4]);
                          [-9, -1, 63]
```

```
> EvalPolyAt([295756, 585756, -101987]);
        [7651205213943063955647, 11772393197745796555647,
                108191226139646787969]
```

3.17   Viewing Maple library source code

As mentioned in the Overview, Maple's kernel performs basic mathematical calculations. This includes: basic integer and polynomial arithmetic, display and file input and output (e.g. most elements of plotting, printing and lprinting, save, read, etc.) and the part of the system that executes input given in Maple's programming language. The kernel is written in the programming language C, and is not available for scrutiny within most versions of Maple.

Maple's library contains most of the mathematical functionality of the system and is viewable. To list a Maple library procedure issue the command interface(verboseproc=2); followed by a print command for the procedure you wish to list. Example 109 shows a linear algebra procedure from the linalg package, and iroot.

Example 109
Displaying the programming for Maple library procedure

Display the procedure to compute the characteristic polynomial of a matrix. This procedure just checks for the proper number of parameters before calling the procedures for the characteristic matrix and determinant.

```
> interface( verboseproc=2 );
> print( linalg[charpoly] );
proc(A,l)
options `Copyright 1990 by the University of Waterloo`;
    if nargs <> 2 then ERROR(`wrong number of parameters`)
    else linalg['det'](linalg['charmat'](A,l))
    fi
end
```

Some procedures consist entirely of programming in C and are not viewable.

```
> print( min );
proc() options builtin; 98 end
```

Maple library procedures that are not automatically loaded should be read in with `readlib` before viewing. `iroot` is a procedure that approximates n^{th} roots.

```
> print( iroot );
                                iroot
> readlib( iroot ):
> print( iroot );
proc(n,r)
options `Copyright 1990 by the University of Waterloo`;
    if type(n,integer) and type(r,integer) then
        if r < 1 then ERROR(`cannot compute root`)
        elif r = 1 then n
        elif n = 1 then 1
        elif n = 0 then 0
        elif n < 0 then
            if irem(r,2) = 1 then -iroot(-n,r) else 0 fi
        else
            trunc(evalf(
                n^(1/r)+1/2,iquo(length(n),r)+4+length(length(n))
                ))
        fi
    elif type(n,numeric) and not type(n,integer) or
         type(r,numeric) and not type(r,integer) then
         ERROR(`arguments must be integers`)
    else 'iroot(n,r)'
    fi
end
```

The master files defining the Maple library's procedures, as written by Maple's developers, contain explanatory comments as well as the procedure definitions. Unfortunately these comments are not viewable by printing the procedure and are not retained when the library procedure is saved or read in. A listing of the Maple library with comments can be obtained for a fee from your Maple distributor.

3.18   Calling Maple from programs written in other languages

Conventional Maple does not provide many facilities for communicating with other programs. The only practical way of using conventional Maple in conjunction with other programming is to establish two programs running concurrently on the computer: Maple, and the rest of the user's programming. Information (expressions, numbers, Maple commands, and so forth) is exchanged through files that Maple and the other program write to and read from each other.[11] This has been done with some degree of success since the early days of Maple development — see, for example, [CMS88].

A special version of Maple also runs on many computers that run conventional Maple. This special version presents Maple as a C language subroutine, that can be called directly from a C (or C-compatible) user program. This "C-callable version" of Maple is more convenient for programmers to use than data exchange through files, and is more efficient. For further details, contact your Maple software distributor.

---

[11]This may be easier on some computers and almost impossible on others. Consult experts for your variety of system to determine the feasibility for doing so on yours.

Chapter 4
Advanced Graphics

In this chapter, we discuss further ways to graph mathematical functions and numerical data. `plot` and `plot3d` provide ways of doing many varieties of two-dimensional and three-dimensional (surface) plotting. The Maple library package `plots` provides convenient ways of doing yet more.

4.1 More on `plot`

`plot( `*expression*`, `*variable*` = `*plot range*`, `*option sequence*` );`	Produce a plot using optional arguments

The simplest forms of plotting were discussed in Section 1.13. The `plot` command can take optional arguments, described in Table 31. In this section, we describe how to apply the options in the situations where you would find them useful.

4.1.1 How `plot` works: adaptive plotting and smooth curves

By default, Maple computes the value of *expression* at a modest number of equally spaced points in the interval. It then selects more points to compute within those subintervals where there is a large amount of fluctuation. This technique is known as *adaptive plotting*, since the number and location of points used by `plot` depends on what it discovers from the information it already has. The adaptive plotting approach does not always lead to "infallible" pictures. Sometimes the picture may differ qualitatively from the actual result. For example, `plot`'s adaptive algorithm does not cope well with many twists and turns in a small area of the curve. Figure 7 illustrates this with a sawtooth function plotted between −1 and 4.

By using the options to plot given in Table 31, you can alter the format of the plot. For example, you can change the number of points used for the initial sample by the adaptive plotting algorithm through the `numpoints` = *n* option to `plot`. Using a value of *n* larger than the default (25 for functions, 49 for parametric plots) often enhances `plot`'s ability to detect sections of the graph that change rapidly within a small interval. You might wish to use this option to achieve a "higher fidelity" picture if you know that the expression you are plotting has such unsmooth characteristics. Figure 8 shows the improvement in the depiction of the sawtooth function that results from using more initial points. However, there are some functions, such as that of Figure 9, which will not be perfectly rendered even when many points are plotted.

Different display devices have differing resolution. For example, a character terminal can typically draw only 80 distinct marks on a row, with 24 rows filling the entire display. Often other displays, for instance of a personal computer or a laser printer, can handle 800 or more distinct marks on a row, using hundreds of rows to form the picture. It would be a waste of effort to try to draw the plot picture using more resolution than the display can provide.

TABLE 31
Options to `plot`

Option	Effect of `plot( ..., option)`
`title` = `Plot title`	The specified title is included in the plot in the upper left quadrant.
`xtickmarks` = n	The x-axis will be drawn with at least n tick marks other than the mark at the origin of the graph. n must be an integer 2 or more. Maple tries to choose a reasonable increment size between tick marks, so you may get a few more tick marks than you specify. If you choose n too large, you can cause the tick mark labels to overwrite each other.
`ytickmarks` = n	The y-axis will be drawn with at least n tick marks. The advice about `xtickmarks` also applies for `ytickmarks`.
`style` = `POINT` `LINE`	The option `style` = `POINT` causes the plotting to be done with points, without lines connecting them. The option `style` = `LINE` causes straight lines to be drawn between the plotted points.
`numpoints` = n	Specifies the minimum number of points to be used in the plot in the adaptive plotting scheme. Without this option, `plot` uses a value of 25 or 49.
`resolution` = n	Inform `plot` that the horizontal resolution of the plot display device is n points. The default value is 200.
`coords` = `polar`	shows that a parametric plot is in polar coordinates. A more convenient method for polar coordinate plotting is to use the `plots[polarplot]` library function. See Section 4.1.7.

The assumption made within `plot` is that the display device has a resolution of 200 distinct elements per row of the plot. You can use the `resolution` option to `plot` to specify otherwise. For example, `plot( ...,` `resolution=400)` would be appropriate if you had a display that has that many elements. The effort would be appropriate because the resolution of the display is much more than the resolution assumed in the default. Most of the time though, you should not need to change the default to see a reasonable looking representation of the graph.

4.1.2   Using titles, lines, and points in your plot

To specify a title on your plot, use `title=`text`` as an option to `plot`. The title usually appears in the upper left quadrant of the plot. Figure 10 is an example of a plot with a title.

By default, `plot` will draw straight line segments between the points it computes for the graph. If you wish to display only the points, give the option `style=POINT` to `plot` (see Figure 12).

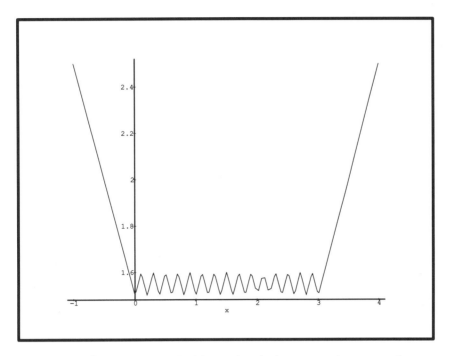

FIGURE 7. `plot(sum((-1)^(i)*abs(x-i/10), i=0..30), x=-1..4);`

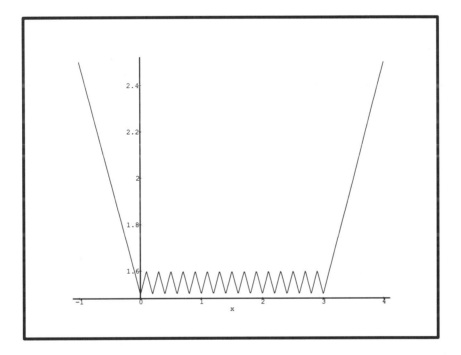

FIGURE 8. `plot(sum((-1)^(i)*abs(x-i/10), i=0..30), x=-1..4,numpoints=500);`

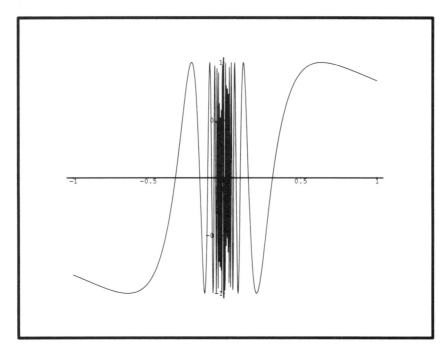

FIGURE 9. Result of `plot(sin(1/x), x=-1..1, numpoints=1000);`

4.1.3   Plotting from a list of data values

`plot( [ [x₁, y₁], ..., [xₙ, yₙ] ], options );`	Plot a list of (x, y) values
`plot( [ x₁, y₁, ..., xₙ, yₙ ], options );`	
`plot( { list₁, ..., listₙ }, options );`	Plot from several lists

You can use `plot` to graph a collection of points, if you know the coordinates of each point through its coordinates (x, y). Collect all the coordinates into a list, starting with the first point's coordinates and continuing without a break to the last point. An alternative to specifying the points in this single lists is to create a list of sublists, with each sublist of giving the x and y values of a point. If you give such a list of points, in either format, as the first argument to `plot`, it will draw a graph with the points connected in the order specified in the list. The "adaptive refinement" described in Section 4.1.1 will not be applied since the fixed information in the list makes refinement impossible. Figure 11 illustrates plotting from a list of points.

Since this mode of plotting allows most of the plotting options of Table 31, you can specify the `style = LINE` or `style = POINT` to override the default. If you specify only one range as an option, then it will be the horizontal range of the plot. A second range will be used as the vertical range. Giving a set of lists will cause several lists of points to be displayed on the same plot.

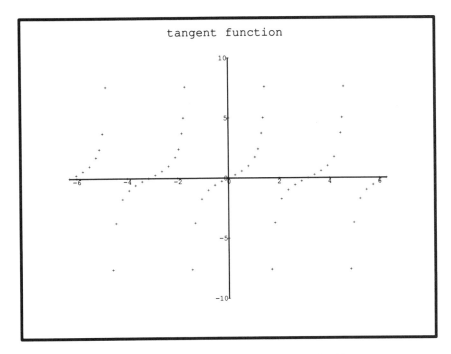

FIGURE 10. Result of `plot(tan(x), x=-2*Pi..2*Pi, -10..10,`
`title='tangent function', style=POINT);`

4.1.4   The `plot` data structure; saving plots into files

When you plot a function or collection of functions, Maple creates a data structure containing the information necessary to draw the picture. The picture is the result of prettyprinting the data structure — a similar idea to printing raised exponents or fractions for mathematical expressions.

The result of `plot` can be assigned as the value to a variable in the same way as other results computed by Maple. If the Maple variable p has as its value a Maple plot data structure (such as through an assignment `p := plot( ...):`), then entering the command `p;` will draw the picture. To see the components of a plot's data structure, `lprint` a plot result, or `lprint` a variable assigned a plot result as its value.

The plot data structure is primarily used for remembering the plotting information for further use. For example, one can assign the result of `plot` to variables and **save** them in files. (This saves the plot information, not the picture.) You can also create new plots by manipulating an existing plot data structure through use of Maple structure operations such as `op`, `subs`, and `subsop`. When **saved** values are **read** into a subsequent Maple session, plots can be redrawn without recalculation of evaluation points by applying `print` or `plots[display]` to them.   See the *Maple V Library Reference Manual* or `?plot,structure` for further information on the `plot` data structure.

4.1.5   Combining plots and zooming in with `display`

Another way that plot data structures can be used is with the `display` function of the `plots` package. If p and q are variables that have plot data structures as their values, then `plots[display]([p,q]);` or `with(plots): display([p,q]);` will draw a picture with both graphs displayed. Plot options can be

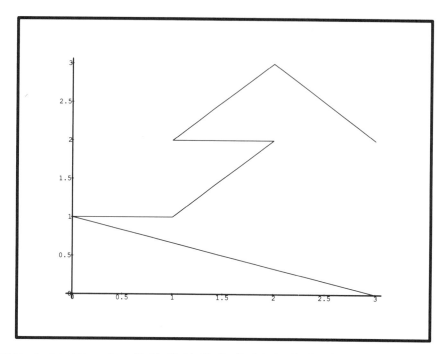

FIGURE 11. A plot of the points $(3, 0)$, $(0, 1)$, $(1, 1)$, $(2, 2,)$, $(1, 2)$, $(2, 3)$, and $(3, 2)$ connected by lines:
```
plot([3,0, 0,1, 1,1, 2,2, 1,2, 2,3, 3,2], style=LINE);
```

given as additional arguments to `display`. By giving a different horizontal and vertical range as options to display, one can "zoom in" on a portion of the plot. `display` uses only the information in the plot data structure provided. Thus "zooming out", while possible, only generates pictures with extra blank space at the ends of the axes. Zooming out does not cause evaluation of the plot expression at any additional points. Example 110 and Figure 12 illustrate the superposition of two plots with a title and vertical/horizontal range options included.

Example 110
Generating a point plot and a plot of a curve; drawing them together via `display`

Here we use an alternative to `plotsetup` that specifies a PostScript format, to be written to the file `display1.ps`.

```
> interface( plotdevice = postscript,
>     plotoutput = `display1.ps` );
> datalist := [ [1950,18.6], [1960,19.3],[1970,20.4],
>               [1980,21.6], [1990,25.4] ]:
```

Create a plot of points, but don't draw it. The display of the plot data structure is suppressed since the line ends with a colon.

```
> PointPlot := plot( datalist, style=POINT,
>   title=`Plot of points`):
```

Look at the plot data structure.

```
> lprint( PointPlot );
PLOT(AXIS(HORIZONTAL,SCALE(4,LINEAR,NUMBER)),AXIS(VERTICAL,
SCALE(4,LINEAR,NUMBER)),CURVE(FUNCTION(LIST([[1950, 18.6],
[1960, 19.3], [1970, 20.4], [1980, 21.6], [1990, 25.4]]))),
BLACK,POINT,[1950., 18.6, 1960., 19.3, 1970., 20.4, 1980.,
21.6, 1990., 25.4]), TITLE(TEXT(Plot of points)))
```

Create an array of data points, plus names of dependent and independent variables.

```
> RegressionInput :=  array( [[t,p], op(datalist)] );
                                   [   t      p   ]
                                   [              ]
                                   [ 1950   18.6  ]
                                   [              ]
                                   [ 1960   19.3  ]
              RegressionInput  :=  [              ]
                                   [ 1970   20.4  ]
                                   [              ]
                                   [ 1980   21.6  ]
                                   [              ]
                                   [ 1990   25.4  ]
```

Create a cubic formula with unknown coefficients a, b, c, d.

```
> cubic :=  a + b*t + c*t^2 + d*t^3;
                                           2       3
                 cubic := a + b t + c t  + d t
```

Find a curve of the variety described by "cubic" that fits the data via least squares regression (from the statistics package).

```
> coefset :=  stats[regression](RegressionInput, p=cubic);
coefset := {a = -389166.3601, b = 601.2581978, c = -.3096383230,
     d = .00005315414143}
```

Generate the formula for the cubic curve.

```
> formula :=  subs(coefset, cubic);
                                                                   2
      formula :=  - 389166.3601 + 601.2581978 t - .3096383230 t
                        3
          + .00005315414143 t
> CurvePlot :=  plot(formula, t=1950..1990):
```

Now superimpose the point plot and curve plot. The option given in display overrides the title of PointPlot.

```
> with(plots):
> display( [PointPlot, CurvePlot], t=1940..2000, p=17..30,
>          title=`Data points and least squares curve fit` );
```

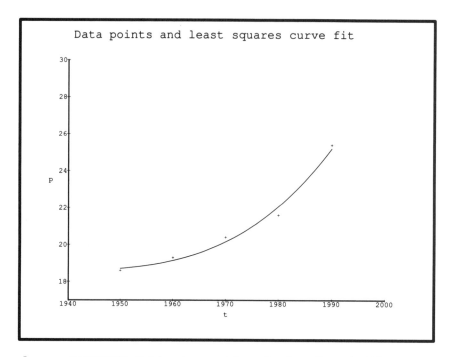

FIGURE 12. Point and curve plot resulting from Example 110

4.1.6   Two-dimensional parameterized plots

```
plot( [x-expression, y-expression, parameter = parameter range] );          Plot of a single function
plot( [x-expression, y-expression, parameter = parameter range],
                    horizontal range, vertical range, options );
plot( { [parameter plot₁], ..., [parameter plotₙ] } );                       Multiple parameter plots
```

In a two-dimensional parameterized plot, two functions, both of a single *parameter*, describe the x and y coordinates of the curve. The curve plotted is from the sequence of (x, y) points resulting from the values of the functions as *parameter* varies through the specified range. As with ordinary two-dimensional plotting, the plot options of Table 31 can be given as extra arguments. Several parameterized plots can be drawn at once by giving a set of parameter plot lists. Figure 13 illustrates parameter plotting.

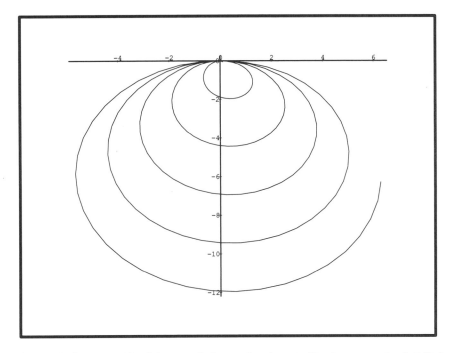

FIGURE 13. A parametric plot: `plot( [t*cos(5*t), sin(5*t)*t-t, t=0..2*Pi] );`

4.1.7   Two-dimensional plots using polar coordinates

`with(plots, polarplot);`	
`polarplot( r-expr, θ-parameter = range, options );`	Plot of a function in polar coordinates
`polarplot( r-expr, options );`	Plot of a function in polar coordinates varying angle between $-\pi$ and π
`polarplot( [r-expr, θ-expr, var = range], options );`	
`polarplot( {polar list_1, ..., polar list_n}, options );`	Polar plot of several functions

A variation of parameterized plotting is to plot using polar coordinates in two dimensions. In the first and second forms of `polarplot` *r-expr* is an expression involving the parameter θ. It describes the radial distance r from the origin $(0, 0)$ as a function of the counterclockwise angle (in radians), from the horizontal axis. (See Figure 14.) You can specify titles, obtain point plots, etc. through the *options* of Table 31. Figure 15 is an example of a plot using polar coordinates.

In the third form of `polarplot`, expressions involving the variable *var* describe the plotted function's r and θ coordinates in parameterized form. As with ordinary parameterized plotting, you can plot several curves on the same graph by giving a set of lists as the first argument to `plot`.

Polar coordinates can often be more convenient for plotting curves that are not easy to describe using the standard coordinate system. For example, you can plot a circle by plotting two semi-circles together using the standard coordinate system:

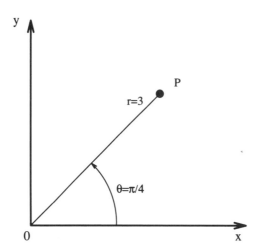

FIGURE 14. Polar coordinates of the point P: $\theta = 45°$ and $r = 3$

```
plot( {sqrt(1-x^2),-sqrt(1-x^2)}, x=-1..1 );
```

or you can do so through a parameter plot:

```
plot( [sin(t),cos(t), t=0..2*Pi] );
```

but the simplest way is to describe the curve in polar coordinates and use the second form of `polarplot`:

```
plots[polarplot](1);
```

4.1.8   Conformal maps

```
with( plots, conformal );
conformal( f );                                        Conformal map of f
conformal( f, z = c_1..c_2 );
conformal( f, z = c_1..c_2, options );
```

`conformal` is useful for visualizing the effect of complex-valued functions. The f argument to `conformal` is an expression involving a single variable z. `conformal` calculates the value of f as z takes on various complex number values. These values will typically be complex numbers themselves, so each will have a

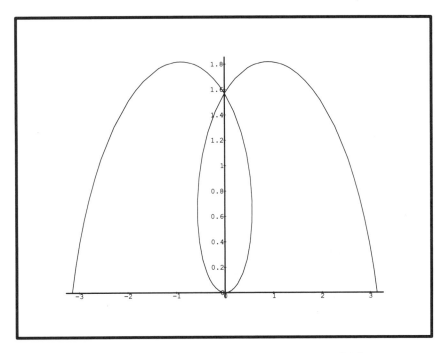

FIGURE 15. A simple polar plot: `plots[polarplot](t);`

real and an imaginary component. We can generate a two-dimensional plot of the complex values, using the real part of the number as the horizontal coordinate and the imaginary part as the vertical coordinate.

`conformal` can help you visualize what happens when z takes on values lying on a grid of equally spaced lines in the complex plane. By default, giving only the f argument to `conformal` causes it to use an 11×11 grid. This grid has its lower left-hand corner at $z = 0$ and its upper right-hand corner at $z = 1 + i$.

Figure 16 illustrates a conformal map of z^2. It is a graph of the results of evaluating z^2 at points taken from the default rectangular grid of complex values.

`conformal` can take several optional arguments described in Table 32. Besides these options, `conformal` can take most of the options of `plot` in Table 31.

4.1.9 A plotting example: a Maple program for simple histograms

Example 111 presents `histogram`, a program to plot simple histograms ("bar graphs") based on a list of integer values. The main procedure `histogram(List, Title)` takes as its argument a list of integer values, and the title to be included in the histogram plot. A key component of the ensemble of procedures is `tower`, which takes as its argument a list $[x, n]$. `tower` constructs a list of points that will become a rectangle of width 1 and height n. The rectangle has its base centered at $(x, 0)$.

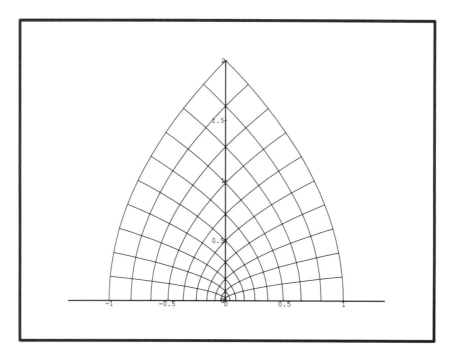

FIGURE 16. Result of `with(plots,conformal); conformal(z^2);`

Example 111
Produce simple histograms of a list of integers

| Given a list `[x,n]`, create a list of coordinates that describe the corners of a rectangle in counter-clockwise order. | ```
> tower := proc(xnlist)
>     local xleft, xright, n;
>     n := xnlist[2];
>     xleft := xnlist[1]-1/2;
>     xright := xnlist[1]+1/2;
>     RETURN([xleft,0,xright,0,xright,n,xleft,n,xleft,0]);
> end:
``` |
```

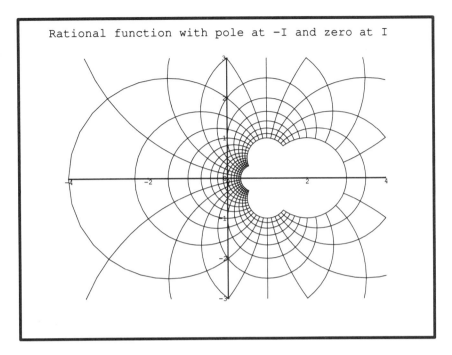

FIGURE 17. Conformal mapping with optional arguments: `conformal( (z-I)/(z+I),`
`z=-2-2*I..2+2*I, -3-3*I..3+3*I, grid=[21,21], numxy=[81,81],`
`title=`rational function with pole at -I and zero at I`);`

zipproc catenates the contents
of two lists into a single list.
histolist is a list of value-count
pairs. plotset is a set of
rectangles for plot3d.
plotstructure is assigned the
plot's data structure. We don't
want a histogram with a
different color for each rectangle,
so we change all the colors to
BLACK by substitution.

```
> histogram := proc( List, Title )
>       local histolist, plotset, count, el, zipproc, palette;
>       count := table(sparse);
>       for el in List do  count[el] := count[el]+1; od;
>       zipproc := (x,y) -> [op(x),op(y)];
>       histolist := zip( zipproc, [indices(count)],
>                                  [entries(count)] );
>       plotset := convert( map(tower,histolist), set ) union
>             { [0, 0, max(op(List))+1, 0] };
>       plotstructure := plot( plotset,      style=LINE,
>                                  title=Title, xtickmarks=8 ):
>       palette := [ RED,   WHITE, YELLOW, GREEN,
>                          BLUE,  CYAN,  MAGENTA ];
>       subs( map(((x) -> x=BLACK), palette), plotstructure );
> end:
> lvals := [1,3,1,7,1,3,7,7,4,6,4,4,2,7,3,3,3]:
> histogram(lvals,`Sample histogram`);
```

histogram uses special features of Maple tables as a programming convenience. A sparse table (see Section 2.10.3) is assigned to histogram's local variable count. This means that entries of count not previously initialized will have value 0. This allows count to keep track of the number of times each integer value in

TABLE 32
Options to `conformal`

Option	Effect of `conformal`(..., *option*)
$d_1..d_2$	When this is the third argument to `conformal`, it specifies a bounding rectangle in the complex plane for the values displayed. For example, in Figure 17, the third argument of `conformal` specifies a bounding rectangle with lower left-hand corner at $-3 - 3i$ and upper right-hand corner at $3 + 3i$.
`grid = [` m, n `]`	Values are chosen from the intersection points of a grid of m horizontal and n vertical, equally spaced lines. `conformal` uses a grid of 11×11 lines without this option.
`numxy = [` i, j `]`	i points are chosen from each vertical line, and j points from each horizontal line. The default is 21 points for each.

`List` occurs without Initializing all its entries. The built-in procedures `indices` and `entries` extract the integers that occurred in `List` with their corresponding counts. Using the built-in procedure `zip` (see Section 2.12) and the "one-line procedure" `zipproc`, the values and count are combined to produce a list of $[x,n]$ pairs. `map` applies `tower` to each component of this list, producing a list of rectanglar coordinates which is then converted to a set. The set's information will be treated as several graphs to be displayed together as the first argument to `plot`.

The other `plot` options specify the title, and that the coordinates of each rectangle must be connected by lines. No picture is generated when the statement containing the `plot` function is executed. Instead, a plot data structure is assigned to the local variable `plotstructure`.

The last two lines of the procedure ensure that all the bars of the histogram come out in the same color (black) on versions of Maple that can display plots in color, such as the version for X Window Systems. The variable `palette` is assigned a list of all the possible colors except `BLACK`. Using `map` to create a list of the form `[RED=BLACK, WHITE=BLACK, ... ]` creates a new plot data structure that substitutes `BLACK` for the other colors.[1] The value of the `histogram` procedure is the plot data structure. When you invoke `histogram` interactively as in Example 111, the result is prettyprinted. Prettyprinting causes display of the histogram. Figure 18 shows plotted result on the sample problem.

This example is meant to illustrate how to achieve graphic effects beyond that provided by the built-in plotting commands. `plots[matrixplot]` with the `heights=HISTOGRAM` is recommended for those who want histograms from Maple in a hurry.

---

[1]There may be many more colors in the palette in future versions of Maple, in which case the approach used here would no longer work. An enlarged palette would come with a built-in library procedure that can determine the colors in a plot.

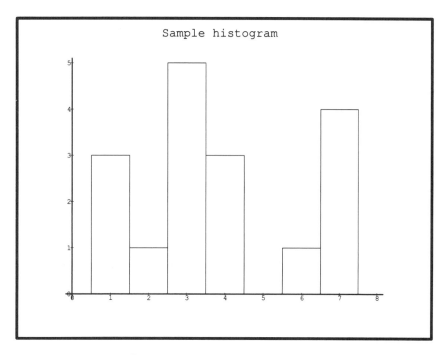

FIGURE 18. PostScript plot of the results of the session of Example 111

4.2   Plotting in three dimensions: graphing surfaces

```
plot3d( f, var₁ = a..b, var₂ = c..d );          Surface plot in three dimensions
plot3d( f, var₁ = a..b, var₂ = c..d, options );
plot3d( set of expressions, var₁ = a..b,        Surface plot of several functions simultaneously
                     var₂ = c..d, options );
```

Maple has a separate command for plotting surfaces in three dimensions. Figure 19 gives an example of such a plot. The expression f to be plotted involves two variables, var_1 and var_2, varying within the ranges $a..b$ and $c..d$ respectively.

plot3d evaluates the expression f on a set of values of var_1 and var_2. plot3d chooses these values automatically from the ranges given for var_1 and var_2. By default, there are 25 equally spaced values used for the first coordinate, and 25 values for the second. The default uses all possible combinations of var_1 and var_2 values: 625 (= 25 × 25) distinct pairs of values.[2]

Each combination leads to a point in three-dimensional space with coordinates corresponding to (var_1 value, var_2 value, f value)[3]. Two points are considered adjacent if their var_1 values are the same and their

---

[2]Unlike plot, plot3d does not use adaptive plotting as discussed in Section 4.1.1. By default, plot3d will use exactly 625 points. The grid or numpoints options of plot3d (see Table 34) change the default.
[3]Conventionally, the first coordinate is x, the second y, and the third z. Maple documentation uses the x-y-z convention on occasion to describe three-dimensional plotting. There is no reason why you can't give the coordinates other names if you wish when you use plot3d.

*var*₂ values are consecutive from those used, or vice versa. By default, `plot3d` will connect adjacent points with lines. This gridwork of criss-crossing lines can be thought of as a surface in three dimensions. It is this surface that `plot3d` depicts in its graphing.

By default, `plot3d` draws the surface using an orthogonal view, that is to say, no transformations for a perspective viewpoint are used. It also hides portions of the surface that would be obscured by other portions of the surface from that viewpoint.[4] People usually draw the x axis in an east-west direction, the y axis in a north-south direction, and the z vertically (see Figure 21). However `plot3d` does not necessarily take this point of view. The point selected is far enough away from the origin to view the entire surface. The viewing angle selected is 45 degrees northeast and 45 degrees above the horizontal plane. This viewpoint can be changed as an option to `plot3d` (see Section 4.2.1).

For most displays for personal computers or workstations, 3D plots can be displayed from a `plot3d` command without preliminary setup. The computer may pause for several seconds after a `plot3d` command before it begins to draw, because a three-dimensional plot often requires significant amounts of calculation. To print 3D plots on PostScript or other printers, proceed as with two-dimensional plotting (see Section 1.13.2, starting on page 28).

`plot3d` supports many options. Some are the same or similar to those offered by `plot`. Others are specific to three-dimensional plots. Tables 33 and 34 give the complete list of options. Subsequent sections of this chapter discuss these options. On many systems with graphical interactive interfaces such as the Macintosh, MS Windows, or X Window System, you can specify these options through menus, dialog boxes, or other means. This can often be more convenient to use than the commands we discuss here. Refer to your interface-specific Maple documentation for further information on these special ways of doing three-dimensional plots.

Display terminals requiring explicit set up before a `plot` (as discussed in Section 1.13.3), also will need the same explicit set up for `plot3d`. Terminals that do not have sufficient resolution to display surface plots, or for which Maple does not have display programming, will cause `plot3d` to issue a message of the form `plot3d is not supported for ... terminals`.

Like `plot`, the result of `plot3d` is a data structure. As with the structures created by `plot`, `plot3d` structures can be assigned to a variable. After assignment, such variables can be **saved**, and **read** into subsequent sessions. The library routine `plots[display3d]`, which works similarly to the `plots[display]` procedure described in Section 4.1.5, can be used to redisplay a data structure, zoom in, add titles, etc. without recomputing the plotted points. Redrawing through `display3d` will be faster than starting over, but there will usually be a pause before the picture appears. This is because the depiction of the surface for the specified viewpoint and options always is computed afresh, though the points defining the surface come from the `plot3d` data structure provided.

4.2.1   Viewing 3D plots: options to 3D plot commands

A photographer makes many choices when taking a picture. These choices include:

- where to position the camera in relation to the subject,
- whether to use "wide angle" or "normal" lenses,

---

[4] "Hidden surface removal" is an important enhancement to the creation of a three-dimensional illusion when drawing. However, there are situations where it detracts from understanding, or where a standard rendition does not help understanding much. You can turn off `plot3d`'s hidden surface removal, or enhance the illusion further by color. See "3D plot styles" in Section 4.2.1.

TABLE 33
Options to `plot3d`

Option	Effect of plot3d(..., *option*)
`title` = `Plot title`	The specified title is included in the plot. (Also see Section 4.1.2.)
`scaling =` UNCONSTRAINED / CONSTRAINED	See "Axes scaling" in Section 4.2.1. The default is UNCONSTRAINED.
`projection` = r	r is a number between 0 and 1. The default is `projection` = 1 (orthogonal perspective). `projection` = 0 is a "wide-angle" perspective.
`view =` $z_{min}..z_{max}$ $[x_{min}..x_{max}, y_{min}..y_{max}, z_{min}..z_{max}]$	See "Viewing boundaries" in Section 4.2.1 for explanation. By default, `plot3d` draws the entire surface.
`orientation` = $[\theta, \phi]$	The viewing angle relative to the origin, in degrees. The default is $\theta = 45$, $\phi = 45$. See "Viewing position" in Section 4.2.1.
`style =` POINT / HIDDEN / PATCH / WIREFRAME	Specifies how a plot is to be drawn. See Section 4.2.5.
`shading =` XYZ / XY / Z	This option works only for displays that use colors or gray shading for the PATCH option. With `shading` = XYZ, color/gray shading varies in all three dimensions. With `shading` = Z, colors vary only in the vertical ("z") direction, staying constant for all points of the plotted surface that are in the same horizontal plane. With XY, color/shading varies in the horizontal plane ("x" and "y" directions), but stays the same for all surface points lying above/below each other. The default shading is device dependent.

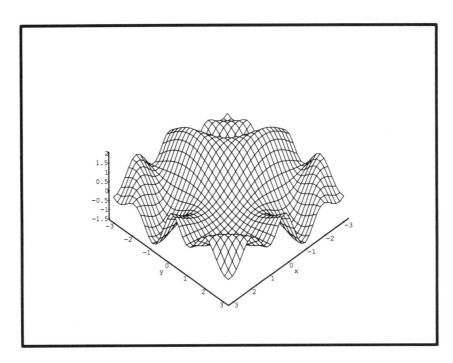

FIGURE 19. plot3d(-exp(-abs(x*y)/10)*(sin(x+y)-cos(x*y)), x=-Pi..Pi, y=-Pi..Pi,
grid=[51,51], style=HIDDEN, scaling=CONSTRAINED, axes=FRAME);

TABLE 34
More options to plot3d

Option	Effect of plot3d(..., *option* = *value*)
axes = BOXED NORMAL FRAME NONE	NORMAL axes are the x and y axes, and the top half of the z axis, all intersecting at the origin. (See Figure 32.) BOXED axes draws a rectangular box enclosing the surface. (See Figure 22.) Three of the edges of the box have x, y and z tickmarks, respectively. The FRAME style only draws the edges of the axes box that have tickmarks in the BOXED style. (See Figure 19.) No axes are drawn if this option is not given, i.e., the default is axes=NONE.
grid = [m , n]	The surface is generated using an $m \times n$ grid of equally spaced values, instead of the default 25×25 grid.
numpoints = n	This is an alternative to the grid method of specifying the number of points. plot3d will use a square grid that includes at least n points.
coords = cartesian spherical cylindrical	Indicates that a parametric plot is in the specified coordinate system. Without this option, plot3d uses cartesian coordinates. See Table 35 for details of sphereplot and cylinderplot, more convenient ways of plotting using spherical or cylindrical coordinates.

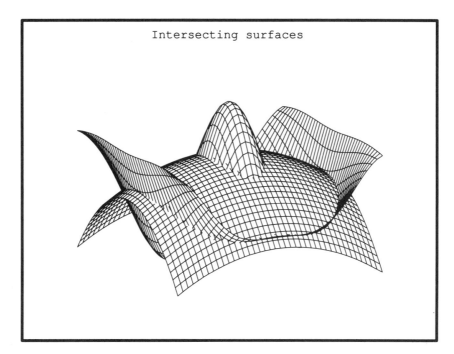

FIGURE 20. `plot3d({cos(sqrt(x^2+3*y^2))/(1+x^2/8), 1/3-(2*x^2+y^2)/19}, x=-3..3, y=-3..3,`
`grid=[41,41], orientation=[-26,71], title=`Intersecting surfaces`);`

- how much of the object to show.

We now discuss the options (summarized in Tables 33 and 34) that provide a way for you to make such choices for your plots.

4.2.2   Viewing position

`orientation = [ `θ`, `ϕ`]`	Select viewing angle, in degrees

As explained in Section 4.2, the default viewpoint is a point $45°$ counterclockwise from the positive x axis toward the positive y axis, and $45°$ "up" toward the positive z axis. These two angles are often called ϕ and θ respectively. Figure 21 illustrates ϕ and θ in the x-y-z axis framework.

One can choose a different viewing angle through the `orientation` option to `plot3d`. In that case `plot3d` chooses a viewpoint with angles for θ and ϕ specified by the option. It selects automatically a position far enough away from the origin so that the entire surface can be seen. Subsequent plotting examples in this chapter, such as Figure 26, illustrate use of this option to specify the viewing angle.

4.2.3   Viewing boundaries

`view =`	$z_{min}..z_{min}$
	`[ `$x_{min}..x_{min}$`, `$y_{min}..y_{min}$`, `$z_{min}..z_{min}$` ]`

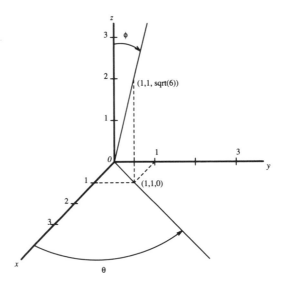

FIGURE 21. ϕ and θ angles in three dimensions

When the `view` option specifies one range of values, `plot3d` will draw only the part of the surface with z coordinates between z_{min} and z_{max}. If a list of three ranges is given, they specify restrictions on the x, y and z coordinates respectively. Figures 22 and 23 illustrate the difference between an unbounded plot and one bounded in z-coordinates.

4.2.4   Axes scaling

`scaling =`	`UNCONSTRAINED`
	`CONSTRAINED`

The figure plotted by `plot3d` lies in a portion of x-y-z space bounded in all three coordinates. We can make independent choices for the scales for all three dimensions using the `view` option previously described in this section. Alternatively we can let `plot3d` make some choices for us. By default (the same as not using the `view` option and using the `scaling = UNCONSTRAINED` option), `plot3d` selects the scaling for all three dimensions so that the figure fills the viewing area/window.

The default (`scaling = UNCONSTRAINED`) usually provides you with a clear view of most of the interesting features of your plotted figure. For example, if the figure's interesting variation in the second coordinate occurs on a much smaller scale than the variation of the first or third, then unconstrained scaling will automatically magnify details of the second coordinate relative to the other two.

With the `scaling = CONSTRAINED` option, all three dimensions use the same ("isotropic") scaling. This might be appropriate when you want a "proportional look". For example, spheres will "look more spherical" under CONSTRAINED scaling. The drawback of constrained scaling (and the strength of unconstrained scaling) is that it may obscure important details when the features in one dimension occur on a much smaller or larger scale than the others. Figures 24 and 25 illustrate the difference between constrained and unconstrained scaling. Figure 19 is another three-dimensional plot that uses constrained scaling.

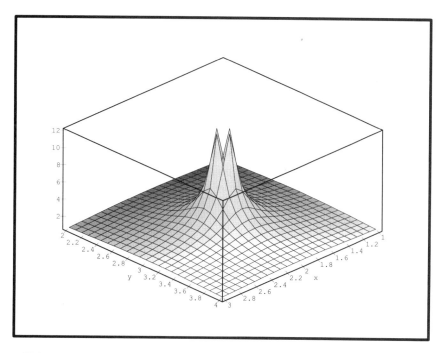

FIGURE 22. Plot of `1/sqrt((x-2)^2 + (y-3)^2)` without viewing boundaries

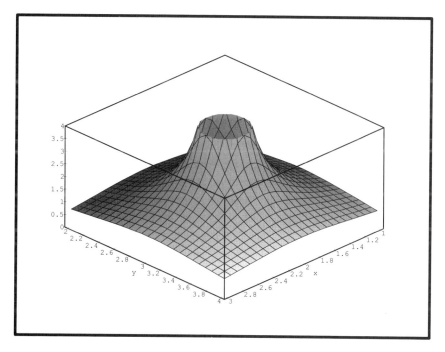

FIGURE 23. Plot of `1/sqrt((x-2)^2 + (y-3)^2)` with viewing boundaries

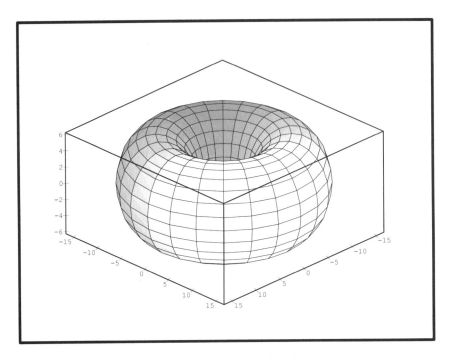

FIGURE 24. Unconstrained plot of a torus

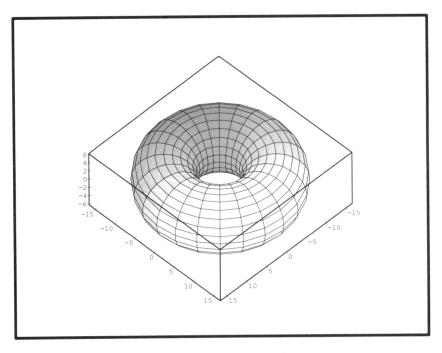

FIGURE 25. Constrained plot of a torus

4.2.5 3D plot styles

By default, Maple draws the plotted surface using only a gridwork of lines. It applies perspective and "hidden-surface removal" so that portions of the surface not visible from the viewpoint are not drawn. Figure 19 illustrates this default.

Other styles are available. For example, you can plot only points without the inclusion of connecting lines, shading, or colors (`style = POINT`).[5] You can draw the gridwork of connecting lines without hidden-surface removal with `style = WIREFRAME` option, or you can specify a "color patch" option (`style = PATCH`). The latter option combines hidden-surface removal with colors or shading to enhance the three-dimensional effect.

Figures 26, 27, and 28 illustrate the options in plotting styles.

4.2.6 Other varieties of three-dimensional plotting

There are a number of other interesting effects you can create with options to `plot3d`, or with procedures from the `plots` package. While we don't have space to describe and illustrate them all, a list of the most useful are in Table 35. See `?plot3d,options ?plots`, the *Maple V Language Reference Manual* and the *Maple*

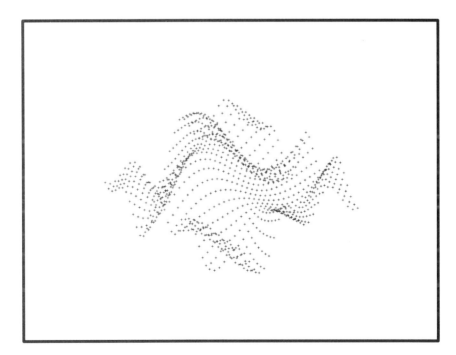

FIGURE 26. Plot of Figure 19 using `style=POINT` option: `plot3d( -exp(-abs(x*y)/10)*(sin(x+y)-cos(x*y))`,
   `x=-Pi..Pi, y=-Pi..Pi, grid=[31,31], style=POINT, scaling=CONSTRAINED, orientation=[-35,55] );`

---

[5]You can draw a list of unrelated points in three dimensions by the `plots[pointplot]` procedure, discussed in Table 35.

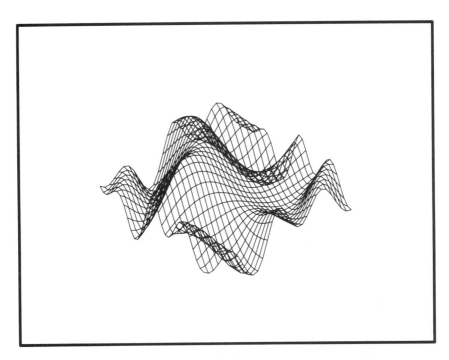

FIGURE 27. Plot of Figure 19 using `style=WIREFRAME` option

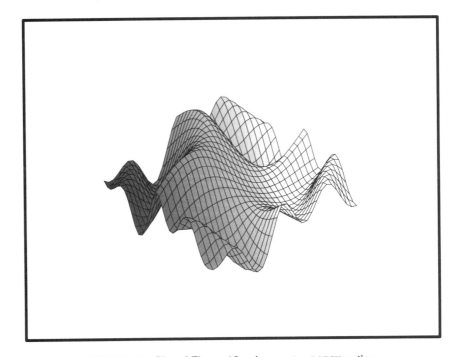

FIGURE 28. Plot of Figure 19 using `style=PATCH` option

V Library Reference Manual for more information. Figures 29, 30, 31, and 3 illustrate some of these effects to whet your appetite for experimentation.

TABLE 35
Other varieties of three-dimensional plotting

```
plot3d( [x-expr, y-expr, z-expr], var₁ = a..b, var₂ = c..d, options );
                                     Parameterized surface plotting. See Figure 29.

plots[sphereplot]( d-expression, θ = angle range₁, φ = angle range₂, options );
                                     Plot using spherical coordinates. See Figure 30.

sphereplot( [d-expression, φ-expression, θ-expression], var₁ = range₁, var₂ = range₂, options );
                                     Parameter plot using spherical coordinates.

plots[cylinderplot]( r-expression, θ = angle range, z = z-range, options );
                                     3D plot using cylindrical coordinates.

cylinderplot( [r-expression, θ-expression, z-expression], var₁ = range₁, var₂ = range₂, options );
                                     Parameter plot using cylindrical coordinates. See Figure 31.

plots[spacecurve]( [x-expr, y-expr, z-expr], var = range, options );
                                     Plot a curve in three dimensions. See Figure 32.

plots[spacecurve]( set of curve lists, var = range, options );          Plot several curves.

plots[tubeplot]( space curve list, var = range );                     Plot a tube in space.

plots[tubeplot]( set of space curves, var = range, options );
                                     Plot several tubes. See Example 1 and Figure 3.

plots[pointplot]( {[x₁, y₁, z₁], ..., [xₙ, yₙ, zₙ]}, options );
                                     Plot a list of points in three dimensions.

plots[pointplot]( [x-expr, y-expr, z-expr], var = range, options );
                                     Plot points taken from a parameterized curve.
```

Example 112
A perturbation of a sphere

```
> interface( plotdevice=postscript, plotoutput=`spherharm.ps` );
```

Generate the Y_7^3 surface spherical harmonic. See [AS65, page 332] for details.

```
> m := 3; n:=7;
```
```
                                   m := 3
                                   n := 7
```

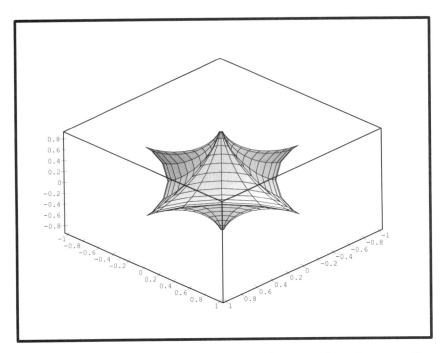

FIGURE 29. 3D parametric plot of $x = \cos^3 v \cos^3 u$, $y = \sin^3 v \cos^3 u$, $z = \sin^3 u$

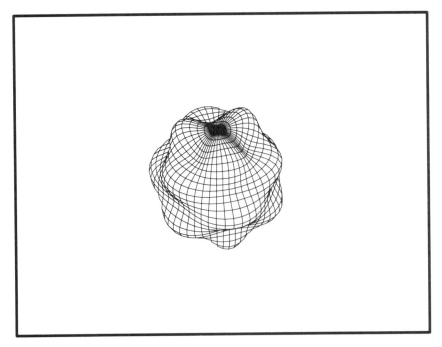

FIGURE 30. Plot using spherical coordinates: A perturbation on the sphere as produced in Example 112.

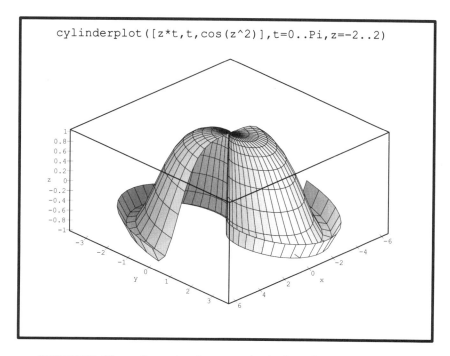

```
cylinderplot([z*t,t,cos(z^2)],t=0..Pi,z=-2..2)
```

FIGURE 31. Three-dimensional parametric plot for cylindrical coordinates

Define the Legendre polynomial $P_n(x)$.

```
> Legendre7 := orthopoly[P](n,x);
```
$$\text{Legendre7} := \frac{429}{16} x^7 - \frac{693}{16} x^5 + \frac{315}{16} x^3 - \frac{35}{16} x$$

Compute the mth derivative. $ is an alternative to seq for constructing expression sequences. x$m constructs a sequence of m xs.

```
> diff(Legendre7 ,x$m);
```
$$45045/8\ x^4 - 10395/4\ x^2 + 945/8$$

Make the substitution $x = \cos(\phi)$ and multiply by an appropriate function of ϕ and θ to calculate the surface harmonic.

```
> Y := subs(x=cos(phi),") * cos(m*theta)*(-sin(phi))^m;
```
$$Y := - (45045/8\ \cos(\text{phi})^4 - 10395/4\ \cos(\text{phi})^2 + 945/8)$$
$$\cos(3\ \text{theta})\ \sin(\text{phi})^3$$

We scale Y to create a small perturbation, on a sphere of radius 5.

```
> plots[sphereplot]( 5+Y/200, theta=0..2*Pi, phi=0..Pi,
>                    grid=[51,31], scaling=CONSTRAINED );
```

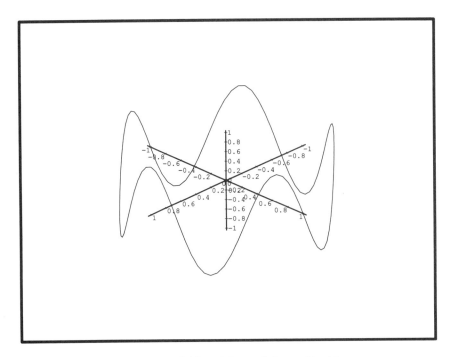

FIGURE 32. plots[spacecurve]([cos(s),sin(s),cos(5*s)],
                     s=0..2*Pi, numpoints=100, axes=NORMAL);

4.3   Plotting functional expressions with `plot` and `plot3d`

plot(*function*, *a..b*)	Two-dimensional plotting
plot([*x-function*, *y-function*, *a..b*])	Parameterized plot
plot({ *functions* }, *a..b*)	Two-dimensional plotting
plot3d(*function*, *a..b*, *c..d*)	Three-dimensional plotting
plot3d([*function₁*,*function₂*,*function₃*], *a..b*, *c..d*)	Three-dimensional parameterized plot

The expressions given to `plot` and `plot3d` implicitly define a function of one or more variables. This function is plotted as the variable or variables vary through a specified range. Alternatively, function name or procedure can be given to the plotting procedures, and the names of the variables omitted. For example, to plot sin on the range $0..2\pi$, one can use an ordinary expression:

```
plot(sin(x), x=0..2*Pi);
```

or just use the function and a range without a variable:

```
plot(sin, 0..2*Pi);
```

Functions that can be plotted in this way include Maple's built-in functions as well as functions that you define and name yourself, such as those described in Section 1.18 or 3.5.

This "functional" approach to plotting can be illustrated by taking up a discussion of the `numeric` option to `dsolve` that uses numerical methods to solve approximately initial value problems involving differential equations instead of finding a formula for the solution.

dsolve({ *differential equation*, *initial condition* }, *dependent function* (*independent variable*), numeric)	Solve a differential equation numerically

`dsolve` with the `numeric` option returns a function that describes a solution to the initial value problem. This result is a Maple procedure requiring one argument. When invoked with a numerical value t, the function returns a sequence of two numbers: t, and the value of the solution at t.

One useful way of employing `dsolve` with the `numeric` option is to assign its result to a variable. This names the Maple procedure that is the solution returned by `dsolve`. You can then plot the function-solution using the functional form of plotting. Since numerical techniques do not deliver a formula describing the solution to the equation, plotting the solution is a good way to understand its characteristics over the domain of interest.

Example 113 and Figure 33 illustrate the use of the functional form of plotting applied to the solution of the numerical form of `dsolve`.

Example 113
Solving a differential equation numerically, and plotting it

Enter a differential equation involving an unknown function y that is a function of a single variable x, and an initial condition for that function.

```
> de :=  diff(y(x),x) = (x-y(x)) / (1+x^2+y(x)^2);
            d                    x - y(x)
      de := ---- y(x) = ---------------
            dx                 2      2
                           1 + x  + y(x)
> ic :=  y(0) = 1;
              ic := y(0) = 1
```

Solve the equation numerically.

```
> f1 := dsolve({de,ic}, y(x), numeric);
f1 := proc(x) `dsolve/numeric/result2`(x,268612020,[1]) end
```

For every numeric argument, `f1` gives a sequence of two values – the argument and the value of y at that argument. The value of y is approximate – established by a numerical differential equation solver.

```
> f1( 0 );
              0, 1.
> f1( 2.5 );
         2.500000000, 1.045905237
```

To plot the solution, define a
function whose value is only the
second value of the sequence.

```
> f :=  x -> f1(x)[2];
                          f := x -> f1(x)[2]
```

Plot it.

```
> plot( f, 0..3 );
```

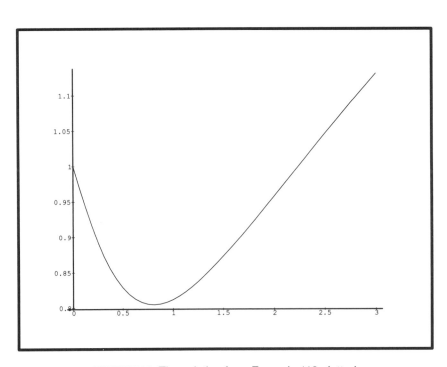

FIGURE 33. The solution from Example 113 plotted

Chapter 5

Measuring and improving performance

5.1 Monitoring time and space consumed during a computation

`time()`	Elapsed time spent in Maple session
`words()`	Memory allocated to Maple data and programs
`words(n)`	Change frequency of memory/time messages

As any Maple user can easily observe, a complicated Maple computation can take seconds, minutes, or more. Symbolic computation can often require a lot of computer memory as well. Computer memory requirements are dictated not only by the length of the answer (the number of terms, number of symbols, or the number of digits in numbers), but also the size of the expressions generated in intermediate steps before the final result. If you are concerned about a Maple computation taking "too long" or "too much memory", the first step to trying to remedy the situation is to find out how much time or memory the computation takes. In this section we discuss ways you can monitor how these resources are being consumed.

The `time` function can be used to measure how long a Maple computation takes. `time` is a function with no arguments. The value of `time()` is how many seconds the computer has spent on Maple computations since the Maple session was started.[1] Taking the difference of two calls to `time` will produce the time elapsed between them. Example 114 illustrates one way to time a Maple computation.

Example 114
Computing elapsed time with the `time` function

Timing a Maple computation on a Sun 4 computer.

```
> time1:=time();
```
$$time1 := .083$$
```
> factor(x*y^2-x);
```
$$x \ (y - 1) \ (y + 1)$$
```
> time2:=time();
```
$$time2 := .166$$

Time of `factor` computation, in seconds.

```
> time2-time1;
```
$$.083$$

---

[1]This feature is not available on all computer systems, since some will not allow timing of programs conveniently. `time` will give a value of 0 on systems where Maple can't time computations.

<table>
<tr><td>Print out number of words of
memory allocated so far.</td><td>

```
> words();
```

31331</td></tr>
</table>

Print out number of words of
memory allocated so far.

```
> words();
```
                                                    31331

Begin printing out bytes used
messages every 50,000 words.

```
> words(50000);
                                    32928
> evalf(Int(sqrt(tan(x)+ln(x^2+1)),x=0..Pi/4));
bytes used=331672, alloc=262096, time=.533
bytes used=539628, alloc=393144, time=.983
bytes used=739692, alloc=589716, time=1.433
bytes used=939912, alloc=720764, time=1.966
bytes used=1140140, alloc=720764, time=2.833
                                    .5689905359
```

Turn off messages.

```
> words(0):
```

The amount of computer memory used for Maple computations is measured in *words* or *bytes*. A byte is 8 binary digits (8 bits). The number of bits in a word of computer memory is typically 32 bits, although some computers are manufactured with a different word size.   All Maple objects (expressions, data structures, and procedures) take up an amount of memory proportional to their size.

The words function describes the amount of computer memory used during a Maple session. The value of words() will be the total number of words of memory used so far in the session.

Some user interfaces for Maple, such as xmaple for X Window Systems displays, show the time and space used within information boxes inside a window. If your version of Maple does not do so, you can use the library showtime function. showtime displays time and space information automatically at the conclusions of every Maple command. It also labels the results computed by Maple O1, O2, etc. The history function just labels results without determining the elapsed time or memory consumed.   See the *Maple V Library Reference Manual* or ?showtime and ?history for further details.

Example 115
showtime and history for timing and labeling results

showtime is a readlib-defined
procedure which labels results
O1, O2, etc. It shows the time
required to complete each
command.

```
> readlib(showtime):
> showtime():
O1 := int(1/(x^2+1),x);
                                    arctan(x)
time    0.08    words    14995
O2 := diff(O1,x);
                                        1
                                      ------
                                        2
                                      x  + 1
time    0.00    words    534
```

Use off to turn off showtime.

```
O3 := off;
```

| The history function is like showtime but it doesn't compute elapsed time. | ```
> readlib(history):
> history():
O1 := solve({x^2-y^2=1,x+y=1},{x,y});
                        {y = 0, x = 1}
``` |
| Note that the labeling has started at O1 all over again. | ```
O2 := subs(O1,x+y);
                              1
``` |
| timing(expr) displays the elapsed time as well as labeling the result. | ```
O3 := timing(int(exp(-x^2)*sqrt(x),x=0..infinity));
                     1/2 GAMMA(3/4)
time = 0.33
O4 := quit
bytes used=487952, alloc=393144, time=.850
``` |

On some versions of Maple, an alternative to using showtime is to have Maple print out messages describing time and memory usage as it computes. These messages are of the form

```
bytes used=xxxx, alloc=yyyy, time=zzzz
```

and are referred to as *"bytes used"* messages. You should not see these messages to begin with since they are turned off initially in typical Maple configurations. An optional integer parameter to words controls the frequency of "bytes used" messages during Maple computations. A command of words(50000); will cause Maple to print out messages at a convenient point after each segment of 50,000 words of memory is used during a Maple session. In general, words(n); will cause time/memory messages to be printed after each increment n words of memory are used. words(0) will turn off the printing of most "bytes used" messages[2]. The words setting can be changed as many times as you wish during a session. On computers where time and space monitoring is built into the display window for Maple, words(n) may change how often the display is updated.

As further explained in Section 5.2, computer memory can be recycled. Thus, while the value xxxx in the "bytes used" message increases with each successive message, the value yyyy of bytes allocated may not always increase. Some memory may have been reclaimed from previous usage.

5.2   Garbage collection and gc

gc()	Perform a garbage collection
gc(*integer*)	Change frequency of automatic garbage collection

---

[2]Changing the frequency of "garbage collection" via the gc function also causes "bytes used" messages to be printed on some systems. If you wish to turn off *all* messages, then you should do both gc(0); and words(0). gc is discussed further in Section 5.2.
```

Some of the expressions generated during a computation are "scratch work" expressions. Although they are necessary, they are never displayed to the user, and need no longer be retained when the final result is produced.[3] The memory used for these intermediate expressions can be reclaimed and recycled for later computations in the session through a process known as *garbage collection*. Garbage collection is important for larger computations because computers do not have an unlimited amount of memory, and it is not unheard of for an ambitious symbolic computation to run through millions of bytes of storage.

During a Maple session, garbage collection will occur automatically according to the pre-programmed rules governing its frequency. The frequency of automatic garbage collection varies from one computer to another. For example, on a typical Sun 4 installation, a garbage collection happens automatically at a convenient point some time after a Maple session has gone through 250,000 words (one million bytes) used, and then again with each succeeding 250,000 increment. You can discover the setting of garbage collection frequency for your copy of Maple by examining the value of the global variable **status**, discussed in Section 5.3.

During a garbage collection, storage no longer being used is reclaimed. Most of the computer memory used by Maple must be scanned to determine what must be retained, and what can be recycled. For longer computations, this can be several million bytes of memory. Thus, in Maple, garbage collections can lead to a noticeable pause in the progress of the computation, from one to several seconds.

The **gc** function can change the frequency and occurrence of garbage collection. **gc()**; will cause a garbage collection immediately, whatever the automatic setting is. **gc(500000)**; will change the frequency of automatic garbage collection so that is occurs with every 500,000 words used. For any positive integer n, **gc(*n*)** will cause an immediate garbage collection, and change the frequency of garbage collection so that it occurs with every *n* increments of words used. **gc(0)** has a special behavior. It does not cause an immediate garbage collection, nor does it change the frequency of garbage collections. It just suppresses printing of "bytes used" messages in subsequent garbage collections.

How often should you have garbage collections occur in Maple? For most work, the default setting should be satisfactory. Typically the frequency of garbage collection is of concern only when running long calculations or in unusual system configurations of memory:

1. When the amount of computer memory allocated is close to the amount of memory you have available, but the amount of memory recycled per collection is significant, consider making garbage collection *more frequent* by decreasing the argument *increment* in **gc(** *increment* **)**. You can discover the amount of memory being returned in a garbage collection by inspecting the value of **status[6]** and **status[7]** — see Section 5.3.

2. When the amount of computer memory available for Maple is many megabytes more than a typical configuration for your type of computer, consider *decreasing* the frequency of garbage collection.

5.3   Querying the state of the system through status

You can discover how much memory or how much time a Maple session has used by inspecting the value of **status**, a global variable. As opposed to **showtime** or **words**, which affect the printing of **bytes used** messages, the value of **status** can be assigned to a variable and referred to in a procedure. **status** is an

---

[3]Even on systems with virtual memory, garbage collection reduces the number of pages needed. This leads to better performance on many virtual memory systems.

expression sequence of 8 values. Table 36 lists the information available through `status`. The value of `status` is updated every time that a `bytes used` message is produced and every time that garbage collection occurs. In between these times it remains unchanged.

TABLE 36
The `status` sequence

Component	Value
status[1]	Total number of words requested from the memory allocator
status[2]	Total amount of memory used by your Maple session (words allocated)
status[3]	Number of seconds of computer time used
status[4]	Increment between `bytes used` messages (in units of computer words), set by default or by `words`
status[5]	Increment between garbage collections (as measured in words), set by default or by `gc`
status[6]	Amount of memory reclaimed in last garbage collection (in words)
status[7]	Amount of memory allocatable for new data (in words, as of last garbage collection)
status[8]	Number of times that garbage collection has been performed since the start of the Maple session

5.4   Profiling the performance of Maple programs

`profile( `*procedure name$_1$*`, `*procedure name$_2$*`, ...)`	Begin profiling procedures
`unprofile( `*procedure name$_1$*`, `*procedure name$_2$*`, ...)`	Turn off profiling
`showprofile()`	Print profile information

If you've determined through simple timing or space measurements that your Maple programs are too slow or take up too much memory, what then? There is usually no substitute for careful attention to what you are doing — substituting an efficient library procedure or implementing a better algorithm yourself. But how can you spend your reprogramming time most productively? It is often the case that most time/memory spent by the computer executing programs is spent within only a small part of the program.[4] If you knew which parts of the program accounted for most of the execution time, you could focus your attention on improving only that portion, saving time and effort.

---

[4]In his famous study of a sample of Fortran programs around Stanford University, Donald Knuth found that "less than 4 per cent of a program generally accounts for more than half of its running time"[Knu71]. This phenomenon — that most of the execution time of a program is spent in only a short part of it — has been found to be true for many other programming languages, including Maple.

Profiling is a way of running experiments with your Maple procedures to indicate where most of the time/memory is spent in executing your actual problems. It can indicate where you can achieve major improvements in performance with a small amount of reprogramming. It can also tell you where effort would be wasted — drastically speeding up a portion of the program that takes only .01% of total execution time to begin with will not help much.

To profile a program properly, you need some "typical test problems" that cause your program to run for several seconds or longer. You also will need some hypotheses about which procedures included in the program are the most expensive. Prepare a file that contains the following:

1. A command to read in all the procedures necessary for the computation, except for built-in library procedures that read themselves in automatically.

2. A `readlib(profile):` command, followed by a `profile` command for the procedures that you suspect of taking the most time.

3. Commands to execute your program on the test problems. You may wish to prepare separate profiles for each test problem, or compute them all as a test suite and collect composite profile data.

4. A `showprofile` command to display profiling information.

5. An `unprofile` command to turn off profiling.

Reading this file into a Maple session will cause execution profiling to be performed on the procedures and subprocedures you have specified. The results of the `showprofile` command should be a table that includes information such as the number of calls to that procedure, the total time spent executing it (and unprofiled procedures called within it), the percentage of the total execution time spent executing it, and the number of words of memory used during execution of it. Using the information from profiling, you can determine which procedures to spend your best efforts tuning. Profiling may also suggest that you look for alternatives that can reduce or eliminate calls to expensive, often-invoked procedures.

Example 116 illustrates the profiling process.

Example 116
An example of program profiling

```
> readlib(profile):
> suc := proc(a) {a} end:
> pre := proc(a) if a={} then {} else op(a) fi end:
> add := proc(a,b) if b={} then a
>  else suc(add(a,pre(b))) fi end:
> mul := proc(a,b) if b={} then {} else
>  add(mul(a,pre(b)),a) fi end:
> pow := proc(a,b) if b={} then suc({}) else
>  mul(pow(a,pre(b)),a) fi end:
> profile(suc,pre,add,mul,pow);
                    suc, pre, add, mul, pow
> pow({{{{}}}},{{{}}});
                        {{{{{{{{{{}}}}}}}}}}
> showprofile();
function       depth      calls         time      time%         words
```

```
----------------------------------------------------------------
suc:
                1       13      0.000   0.0000%              875
pre:
                1       20      0.017   9.2896%             1277
pow:
                1        1      0.167  91.2568%            16728
                2        1      0.050  27.3224%             6271
                3        1      0.000   0.0000%              421
add:
                1        6      0.100  54.6448%             9845
                2        6      0.066  36.0656%             5163
                3        3      0.017   9.2896%             2585
                4        3      0.000   0.0000%              178
mul:
                1        2      0.150  81.9672%            14477
                2        2      0.101  55.1913%             9672
                3        2      0.050  27.3224%             4965
                4        2      0.000   0.0000%              131
----------------------------------------------------------------
total:                         0.183                      20393
> unprofile(suc,pre,add,mul,pow);
                 suc, pre, add, mul, pow
```

5.5   Using option remember to improve performance

In Section 3.10, option remember was introduced as a way to avoid recomputation by using an efficient way of looking up previously computed results. option remember can in certain situations lead to dramatic speedups. One type of computation where option remember can lead to dramatic results are those defined by *recurrence relations* — where $f(n)$ is defined in terms of $f(n-1)$, $f(n-2)$, etc. By using option remember in the programming for f, one can save considerable time by avoiding recomputation. For example, in the computation of the Chebyshev polynomials of Example 95, the computation of Chebyshev(10,_XX) involves recursive calls to Chebyshev2(_XX,9), Chebyshev2(_XX,8), ... Chebyshev2(_XX,0), but only one call each. Thus a total of 10 calls to procedure Chebyshev2 are involved, with another 10 "remember table" lookups.[5] If Chebyshev2 is programmed without option remember then 177 calls to Chebyshev2 are required! While there is no substitute for actually *thinking* about the algorithm you are using and seeing if there is a much better alternative available, option remember is expedient and can often produce good results.

While "remembering" is often a good idea in highly recursive or frequently called functions, it is not a panacea, nor always the way to the best performance. A program to compute the nth Fibonacci number through the recurrence $f_n = f_{n-1} + f_{n-2}$ with $f_1 = 1$, $f_0 = 1$ would improve its performance dramatically through option remember. But Maple's combinat[fibonacci] library function uses a better idea: the observation that for $A = \begin{pmatrix} 1 & 1 \\ 1 & 0 \end{pmatrix}$, $\begin{pmatrix} f_{n+1} \\ f_n \end{pmatrix} = A^n \begin{pmatrix} f_1 \\ f_0 \end{pmatrix}$ so that f_n can be computed by computing the

---

[5]Remember table lookups usually proceed much more quickly than procedure invocations. But the key factor in speeding performance is avoiding a long series of recomputations of Chebyshev2 for various values of its second argument.

matrix power A^n efficiently. (See "Lookup Tables, Recurrences and Complexity" by R. Fateman[Fat89] for a further discussion.)

Although `option remember` seems like a good idea, it does have a cost — the memory used to remember previously computed results. Indiscriminate use of it in your programming can lead to degradation of performance or "out of memory" messages. If you make extensive use of remembering in large computations, it can be cumbersome to erase remembered information explicitly through `forget` (discussed in Section 3.10). If a procedure has `option system` as well as `remember` in its definition the remember table is purged of all unreferenced entries during every garbage collection. For more information, see "Remember Tables" in the *Maple V Language Reference Manual*, or `?remember`.

5.6   Faster floating-point computation

5.6.1   Maple's hardware floating-point arithmetic is faster than `evalf`

Programming languages such as Fortran typically use the computer's floating-point circuitry which has a fixed number of digits of precision. Maple's floating-point arithmetic as used by `evalf` or `fsolve` uses the computer's integer arithmetic. The `evalf` approach to floating-point arithmetic makes it easy to change `Digits` of precision or to use a large number of digits, but it sacrifices speed compared to floating-point circuitry (also known as *floating-point hardware*).

There is another way that `evalf` floating-point arithmetic can be slower than floating-point arithmetic done through other languages. Pre-translation from the user's programming language into the machine's own instruction language (known as *compilation*) can speed up execution over Maple's way of executing programs (known as *interpretation*). With compilation, the translation overhead is incurred once, before the program is executed. With interpretation, there is translation overhead throughout program execution.

It makes sense to compare Maple's interpreted `evalf` floating-point arithmetic with that of a program written in a language such as Fortran or C where both compilation and hardware floating-point arithmetic are used. To be fair, we should use a value for `Digits` in Maple that corresponds to the accuracy provided by hardware floating-point arithmetic. Tests (see the *Maple V Library Reference Manual*, page 84) suggest that for comparable computations, `evalf` can be 50 to 500 times slower than the equivalent Fortran program. When does this drastic difference in speed matter? On a typical workstation or personal computer, calculations requiring a few hundred or even a few thousand arithmetic operations will take a only a few seconds time with `evalf`. Since it is usually much faster to type in an `fsolve` or `evalf` command than to write and compile a conventional program, user convenience gives a performance edge over raw execution speed for such situations. Only for longer computations involving numbers with modest numbers of `Digits` does speed in arithmetic begin to be a serious performance issue. Maple provides an alternative to `evalf` called `evalhf` that executes Maple procedures using hardware floating-point arithmetic. We discuss this performance booster in the next sections.

5.6.2   `evalhf` uses hardware floating-point arithmetic

`evalhf( ` *expression* ` )`	Evaluate using hardware floating-point

`evalhf` works like `evalf` in that all numerical operations are performed using floating-point arithmetic. Like `evalf`, Maple includes built-in programming that allows automatic conversion and approximation of symbolic constants such as π or γ or mathematical functions such as sin or ln. On most computers, `evalhf` will use 15 decimal digit (64 bit) floating-point arithmetic, since that is the kind usually supplied in floating-point hardware.[6] Only operations that produce (floating-point) numerical results are allowed. However, `evalhf` is more restrictive than `evalf` in the types of computations you can perform within it.

When you evaluate *expression* through `evalhf`, all arguments are converted to hardware floating-point numbers. With a few exceptions the only operations permitted during `evalhf` evaluation of *expression* are those that result in numerical results. Symbolic or structural operations such as those performed by `op`, `solve`, or `normal` result in errors in `evalhf`. When evaluation is complete, the hardware floating-point result is converted back to a standard Maple floating-point number (the kind described in Section 1.12) before any assignment or printing is done by Maple. Thus `evalhf` users have only indirect access to hardware floating-point numbers. Example 117 illustrates a mathematical expression evaluated through `evalhf`.

Example 117
Speeding up floating-point arithmetic with `evalhf`

`evalhf` treats the symbol `Digits` in a special way – it doesn't consult its value, it returns the number of digits that `evalhf` uses. Hardware floating-point arithmetic typically uses binary arithmetic. On the computer where this example is being run we have about 15 decimal digits.

```
> Digits;
                              10
> evalhf(Digits);
                             15.
```

Label each result and show the time spent computing it.

```
> readlib(showtime):
> on();
```

Time the computation of an expression via `evalf` and `evalhf`. `evalhf` is so fast that the timer says that it took less than .01 seconds.

```
O1 := evalf(ln(cos(1/2)^(3.5)*exp(Pi)));
                    2.684547812
time   0.05   words   8269
O2 := evalhf(ln(cos(1/2)^(3.5)*exp(Pi)));
                  2.684547812036764
time   0.00   words   149
```

[6]Some personal computers do not have hardware support for "fast" floating-point arithmetic, such as an Apple Macintosh without an FPU (floating-point unit). A Macintosh without an FPU will use an integer/software implementation of floating-point arithmetic similar to Maple's own.

Simple procedures involving
only the computation of numeric
quantities can be evaluated
using evalhf.

```
03 := f := proc(n)
03 :=      if n<2 then RETURN( n )
03 :=      else RETURN( ln(n+1) * f(n-1) / n )
03 :=      fi;
03 := end;
proc(n)
    if n < 2 then RETURN(n) else RETURN(ln(n+1)*f(n-1)/n) fi
end
time   0.01   words   386
04 := evalf( f(100) + f(10) + f(1) );
                        1.000059413
time   0.75   words   67647
05 := evalhf( f(100) + f(10) + f(1) );
                        1.000059412887470
time   0.03   words   193
06 := quit
bytes used=413252, alloc=327620, time=.966
```

5.6.3   Taking advantage of evalhf in your Maple programming

evalhf expects that every operation involves floating-point numbers, other than for a few exceptions to be explained in this section. If you have written a procedure where all operations are numerical, then you can evaluate this procedure through evalhf and achieve the floating-point arithmetic efficiency gains discussed in Section 5.6.1.

Once evaluation under evalhf begins, numbers given to it are converted to hardware floating-point numbers and all further operations are performed using hardware floating-point arithmetic. If any intermediate result or any operand is non-numeric, evalhf gives an error. Thus if you wish to execute your procedure through evalhf, it cannot contain operations on sets, lists, symbolic formulae, etc. Arrays of floating-point numbers are permitted as an exception to this rule, as discussed later in this section.

Exceptions to these general rules make this Spartan (or at least Fortran-esque) regime of programming tolerable:

1. Expressions that evaluate to true or false under ordinary Maple rules are permitted. But outside of if or while statements, 1.0 is used for true and 0.0 for false. Equations and inequalities will be treated as if an evalb had been applied to them, to force them to a numerical result.

2. Under evalhf, type(*expression*, numeric), type(*expression*, float), and type(*expression*, rational) will always evaluate to true as long as the value of *expression* is a floating-point number. type(*expression*, integer) will evaluate to true if the value of *expression* is a (floating-point) integer.

3. Arrays can be passed in as arguments to procedures evaluated under evalhf, and arrays of floating-point results can be returned. If you wish to use an array A to pass values into a procedure, just use it as an argument to the procedure as under ordinary Maple rules. If you wish to use an array A both as a way to import values into a procedure and as a way of exporting results, you must tag the array using var(A) instead of just A. An array passed into evalhf without a var tag does not get altered by the actions within evalhf. The act of converting to hardware floating-point format creates a copy of an array; these copies are altered within evalhf but only results for array arguments with a var tag get copied back to the original array. See Example 118 for an illustration of array tagging.

Other notable differences in the handling of arrays under `evalhf`:

(a) Unassigned array elements are treated as if they had been initialized to `0.0`.

(b) There is no way to determine the dimension of an array `A` within a procedure through expressions such as `linalg[vectdim]` or `op(1,A)`. You should either pass in the dimensions as separate parameters, or, if the array is created within a procedure, to remember the dimension(s) in variables.

(c) References to global array variables generate an error within `evalhf`.

(d) Arrays cannot use special indexing functions such as `symmetric` (see Section 2.10.2), nor can arrays created within a procedure executed under `evalhf` use the initialization feature of the `array` procedure. This is mentioned in Section 2.10.3.

As Example 117 illustrates, `evalhf` treats the global variable `Digits` in a special way — its value under `evalhf` is the number of decimal digits used by hardware floating-point arithmetic instead of whatever value it has been assigned. Access to other characteristics of the floating-point arithmetic used by `evalhf` are available by inspecting the values of the special global variables of Table 37, which are the defining parameters of the widely used IEEE floating-point arithmetic standard.

Using these values it is possible in principle to write Maple procedures that adjust themselves to work with maximal accuracy provided by the hardware floating-point arithmetic of the computer they are executed on, even if this differs from what is typical. Prior experience, expertise in numerical analysis, and careful thought is usually necessary to write such self-adjusting routines correctly, however.

TABLE 37
Global variables with special values under `evalhf`

FLT_RADIX	DBL_MANT_DIG	DBL_DIG	DBL_EPSILON
DBL_MIN_EXP	DBL_MIN	DBL_MIN_10_EXP	DBL_MAX_EXP
DBL_MAX	DBL_MAX_10_EXP	LNMAXFLOAT	Digits

Example 118
Using arrays of floating-point numbers with `var` and `evalhf`

Define a procedure to multiply $n \times n$ matrices A and B, placing the result in C.

```
> matmul := proc(n, a, b, c)
> local i, j, k, s;
> for i to n do
>  for j to n do
>    s := 0;
>    for k to n do
>      s := s + a[i,k] * b[k,j];
>    od;
>    c[i,j] := s;
>  od;
> od;
> n
> end:
```

Set up Maple floating-point computation to use the same number of digits as `evalhf`. Choose $n = 30$ as the size of the test matrices to be generated. Each entry of the test matrices A and B will be a randomly generated floating-point number.

```
> Digits := trunc(evalhf(DBL_DIG));
                                Digits := 15
> N := 30:
> r2 := rand(1-10^Digits..10^Digits-1):
> r := proc() Float( r2(), -Digits ) end:
> A := array(1..N,1..N):
> B := array(1..N,1..N):
> C := array(1..N,1..N):
> for i to N do
>  for j to N do
>    A[i,j] := r();
>    B[i,j] := r();
>    C[i,j] := 0;
>  od;
> od:
```

Time the `evalhf` computation.

```
> st := time():
> evalhf( matmul(N, var(A), var(B), var(C)) ):
> evalhf_time := time()-st;
                        evalhf_time := 2.516
```

Compare this to the time of the ordinary Maple floating-point computation.

```
> st := time():
> matmul(N, A, B, C):
> evalf_time := time()-st;
                        evalf_time := 22.067
> print( `The speedup factor is approximately`,
>      round(evalf_time/evalhf_time) );
             The speedup factor is approximately, 9
```

5.6.4   evalhf is still slower than compiled floating-point programs

Although `evalhf` improves Maple's performance on fixed-precision floating-point computation, languages like Fortran still hold a performance advantage because of compilation. `evalhf` computations can be between 5 and 50 times slower than the equivalent Fortran program. Nevertheless, `evalhf` offers a straightforward way to combine the convenience of Maple with the high performance of hardware floating-point arithmetic. It may be the fastest way to obtain your results. For very extensive floating-point calculations, one might consider developing the computation in Maple and testing it on modest-sized problems, then using the `fortran` or C translation facilities described in Section 2.18 to convert the programming to a compiled language supporting "fast" floating-point.

   A portion of Maple that is particularly floating-point intensive is plotting. Both `plot` or `plot3d` use hardware floating-point arithmetic when it is available. You do not need to use `evalhf` when plotting.

Chapter 6
Advanced Examples

6.1 Introduction

This chapter shows Maple in action on a collection of problems drawn from engineering, mathematics, and the physical sciences. Here's a list of the examples and the Maple techniques they illustrate:

Balancing chemical reactions (Section 6.2): Equation solving with `isolve`; back substitution of a solution into the defining equations.

The Maxwell gas velocity formula (Section 6.3): Finding a maximum of a function through `diff` and `solve`; simplification of algebraic expressions with `simplify`.

The critical length of a rod (Section 6.4): Differential equations solved through `dsolve`; name aliasing; substitution of identities; one-sided limits; two-dimensional plotting.

Zeros of Bessel functions (Section 6.5): Name aliasing; constrained root-finding with `fsolve`; plots of several functions; formatted output of a table.

Stock market computation (Section 6.6): Linear algebra with `linalg`; matrices; eigenvalues and eigenvectors; characteristic matrices and null spaces.

Primitive trinomials (Section 6.7): Modular integer and polynomial arithmetic; factorization of integers and polynomials (`ifactor` and `factor`); factorization over finite fields; iterated computation using `for-in-while`.

The $3n + 1$ conjecture (Section 6.8): Number theory; large integers; type checking with structured types; selection of a result through `op`; integer quotients and remainders (`iquo`); definition of a recursive function through `proc` and through assignment.

Numerical approximation of a function (Section 6.9): Automatic integration through `evalf`; Taylor series computation; function definition through `unapply`; rational function approximation with `convert(...,ratpoly)`; Chebyshev polynomials, Chebyshev approximation with `chebyshev`.

6.2 Balancing chemical reactions

In this example, we find a solution for balancing a chemical reaction for the equation

$$a\,\mathrm{Cu_2S} + b\,\mathrm{H^+} + c\,\mathrm{NO_3^-} \rightarrow d\,\mathrm{Cu^{2+}} + e\,\mathrm{NO} + f\,\mathrm{S_8} + g\,\mathrm{H_2O}$$

Each element is a conserved quantity. We enter a labeled equation for each element.

```
> Cu := 2*a = d:
> S  := a   = 8*f:
> H  := b   = 2*g:
> N  := c   = e:
> O  := 3*c = e+g:
```

The electrical charge Q must also be conserved.

```
> Q := b-c = 2*d:
```

We wish to find the smallest solution possible for *integer* values for a through g. Maple's `isolve` function will do exactly this for us.

```
> ans := isolve( {Cu,S,H,N,O,Q} );
     ans := {a = 24 _N1, d = 48 _N1, f = 3 _N1, b = 128 _N1, g = 64 _N1,

        c = 32 _N1, e = 32 _N1}
```

The result includes a parameter _N1 for which we are free to choose any integer value. We set _N1 to 1 to get the simplest solution.

```
> subs(_N1=1,ans);
        {a = 24, g = 64, b = 128, f = 3, d = 48, c = 32, e = 32}
```

6.3   Maxwell's formula for the velocity of a gas sample

In this example, we solve a small problem in statistical mechanics, namely finding the root mean square (rms) velocity of a gas sample. According to the Maxwell-Boltzman theory, the velocity distribution given by the following formula specifies the likelihood of a molecule in the sample having speed v.

```
> n := v -> 4*Pi*(n0)*( (m/(2*Pi*k*T))^(3/2) )*exp(-m*v^2/(2*k*T))*v^2:
> n(v);
```

$$
\frac{n0\ 2^{1/2}\ m^{3/2}\ \exp\!\left(-\frac{1}{2}\dfrac{m\,v^{2}}{k\,T}\right)v^{2}}{Pi^{1/2}\ k^{3/2}\ T^{3/2}}
$$

where k, m, and T represent the following:

T	-	absolute temperature
m	-	molecular mass
k	-	Boltzman's constant

The problem is to determine the most likely speed of a particle. This is obtained by finding the speed v such that $n(v)$ is maximized. This is the point where the derivative of $n(v)$ is zero.

```
> dv := diff(n(v), v);
```

$$
dv := -\frac{n0\ 2^{1/2}\ m^{5/2}\ v^{3}\ \exp\!\left(-\frac{1}{2}\dfrac{m\,v^{2}}{k\,T}\right)}{Pi^{1/2}\ k^{5/2}\ T^{5/2}} + 2\,\frac{n0\ 2^{1/2}\ m^{3/2}\ \exp\!\left(-\frac{1}{2}\dfrac{m\,v^{2}}{k\,T}\right)v}{Pi^{1/2}\ k^{3/2}\ T^{3/2}}
$$

```
> CritPoints := solve(dv = 0, v);
```

$$
CritPoints := 0,\ \frac{2^{1/2}\ k^{1/2}\ T^{1/2}}{m^{1/2}},\ -\frac{2^{1/2}\ k^{1/2}\ T^{1/2}}{m^{1/2}}
$$

The third value is clearly not in the running for the maximum since it is negative.

Now we use the second derivative test to find the point of maximum.

```
> test1 := simplify( subs(v = CritPoints[1], diff(n(v), v$2)) );
```

$$
test1 := 2\,\frac{n0\ 2^{1/2}\ m^{3/2}}{Pi^{1/2}\ k^{3/2}\ T^{3/2}}
$$

```
> test2 := simplify( subs(v = CritPoints[2], diff(n(v), v$2)) );
                             1/2  3/2
                        n0  2     m     exp(-1)
            test2 := - 4 -------------------
                             1/2  3/2  3/2
                           Pi    k     T
```

The point where the second derivative is negative is the point of maximum.

```
> maxv := CritPoints[2];
                              1/2  1/2  1/2
                             2    k    T
                    maxv := --------------
                                 1/2
                                m
```

The value `maxv` is the well-known *rms* velocity.

6.4   Critical length of a rod

This example will illustrate how a combination of symbolic math techniques, numerical methods, and graphics can be used together to solve a problem. This problem of finding the critical length of a rod is an engineering problem adapted from *Advanced Engineering Mathematics* by Peter O'Neil[O'N91].

Suppose that we have a thin elastic rod standing vertically with its bottom end embedded in a slab of concrete. The rod is given a small push horizontally at its top end.

- If the rod is *short* enough, it will *spring* back after being pushed

- If the rod is *long* enough, it will *stay bent* after being pushed

The critical length is the cross-over point and it is determined by the thickness of the rod, its weight, and the elasticity of the material used for the rod. We will use the following symbols to represent these parameters:

L	-	Length of the Rod
A	-	Area of circular cross-section
M	-	Young's modulus (elasticity)
W	-	Weight per unit length

From the theory of elasticity, we can model this problem by the third order differential equation:

$$\frac{d^3 y}{dx^3} + \frac{W}{MA} x \frac{dy}{dx} = 0$$

where x varies from 0 at the rod's base to L, its length at the top of the rod and $y(x)$ is the horizontal displacement along the rod. Details on the derivation of this differential equation are given by O'Neil.

By a simple substitution, we can derive a second order equation. Let $u = y'$ to get the following equation:

$$\frac{d^2 u}{dx^2} + \frac{W}{MA} x u = 0$$

As we enter this equation into Maple, we will use C to represent the constant expression $W/(M * A)$.

```
> diff(u(x),x,x) + C*x*u(x) = 0;
                / 2     \
                | d     |
                |----- u(x)| + C x u(x) = 0
                |  2    |
                \ dx    /
```

Now we can use the `dsolve` command to solve this differential equation :

```
> dsolve( ", u(x) );
        1/2                    1/2 3/2                        1/2 3/2
u(x) = x    (_C1 BesselJ(1/3, 2/3 C    x   ) + _C2 BesselY(1/3, 2/3 C    x   ))
```

Note that arbitrary constants _C1 and _C2 were introduced since boundary values were not given for the function u. These names which are generated automatically by Maple start with the underscore character to avoid any conflicts with any names which you may already be using.

We will take advantage of Maple's `alias` facility to introduce abbreviations for the names `BesselJ` and `BesselY`. The names `BesselJ` and `BesselY` will still be used internally in Maple computations but aliasing will allow us to use J and Y, respectively, as abbreviations. Maple will also translate instances of the longer names to the shorter abbreviations upon output.

```
> alias( J=BesselJ, Y=BesselY);
                        I, J, Y
```

The list of names returned by `alias` shows the list of alias names that are currently defined. The list of aliases that Maple uses initially contains the alias I=(-1)^(1/2). Since $2/3 * C^{1/2} * x^{3/2}$ occurs several times in the uses of the Bessel functions above, let's introduce one more abbreviation.

```
> alias( X = 2/3*C^(1/2)*x^(3/2) );
                           I, J, Y, X
```

We will assign the right-hand side of the defining equation for u, which occured three expressions earlier, to the variable `solution`.

```
> solution := rhs("""");
                       1/2
           solution := x    (_C1 J(1/3, X) + _C2 Y(1/3, X))
```

Instances of the Bessel Y function can be replaced by a combination of Bessel J functions through the identity

$$Y_\nu(x) = \frac{1}{\sin \nu\pi}[J_\nu(x)(\cos \nu\pi) - J_{-\nu}(x)]$$

We can use `subs` to apply this rule to the value of `solution`.

```
> subs( Y(1/3, X ) = 1/sin(Pi/3) * (J(1/3, X) * cos(Pi/3) - J(-1/3, X)),
>    solution );
         1/2                        1/2
        x    (_C1 J(1/3, X) + 2/3 _C2 3    (1/2 J(1/3, X) - J(-1/3, X)))
```

We can see that the constant expressions can be folded together and replaced by new constants $c1$ and $c2$ to get a simpler expression.

```
> solution := c1 * x^(1/2) * J(1/3,  X) + c2 * x^(1/2) * J(-1/3, X);
                       1/2              1/2
           solution := c1 x    J(1/3, X) + c2 x    J(-1/3, X)
```

At the upper end of the rod, we know that the bending moment is 0. Therefore,

$$\left.\frac{d^2y}{dx^2}\right|_{x=0} = 0$$

Since $u = y'$, we need only differentiate the expression for $u(x)$ once with respect to x to obtain the second derivative of y.

```
> diff(",x);
      c1 J(1/3, X)            /                   J(1/3, X)\  1/2           c2 J(-1/3, X)
1/2 ------------ + c1 x |- J(4/3, X) + 1/2 ---------| C     + 1/2 -------------
        1/2             |                    1/2 3/2|                       1/2
       x                 \                   C    x  /                    x


         /                   J(-1/3, X)\  1/2
 + c2 x |- J(2/3, X) - 1/2 ----------| C
        |                    1/2 3/2|
         \                  C    x   /
```

The denominators in this answer contain powers of x, which prevents us from simply evaluating this expression at $x = 0$. However, we can find the limit of this expression as x approaches 0 from the right.

```
> limit( ", x=0, right );
                          1/6  1/6
                      c1 3    C     GAMMA(2/3)
                  3/2 ----------------------
                             Pi
```

This expression is the bending moment at the upper end of the rod, which we know to be 0. Therefore $c1$ must be 0. Let's substitute 0 for $c1$ in the definition for $u(x)$.

```
> subs( c1=0, solution );
                      1/2
                  c2 x     J(-1/3, X)
```

Since the lower end of the rod is fixed and cannot move

$$\left.\frac{dy}{dx}\right|_{x=L} = 0.$$

We know that $u = y'$ evaluated at L should be 0.

```
> subs( x=L, " );
                  1/2                   1/2 3/2
              c2 L     J(-1/3, 2/3 C    L   )
```

Unless $c2$ is zero, the value of the Bessel function with the argument given must be 0. Therefore $2/3 C^{1/2} L^{3/2}$ must be a root of $J_{-1/3}(x) = 0$.

```
> firstRoot = 2/3 * C^(1/2) * L^(3/2);
                                            1/2  3/2
                      firstRoot = 2/3 C    L
```

and *L* can be defined as

```
> solution := L = ( firstRoot / (2/3*C^(1/2)) ) ^ (2/3);
                                2/3  1/3            2/3
                               3    2     firstRoot
                solution := L = 1/2 ----------------------
                                             1/3
                                            C
```

To get an idea of where the first positive root is, let's take a look at the function near zero (Figure 34):

```
> plot( J( -1/3, x ), x=0..5 );
```

We see that the first root occurs between 1 and 2. Now let's use numeric root-finding techniques to locate the root in that range:

```
> fsolve( BesselJ(-1/3,x), x, 1..2 );
                           1.866350859
```

Substitute the root into the equation for *L*:

```
> subs(firstRoot=1.866350859, solution);
                                   2/3  1/3
                                  3    2
                   L = .7579354115 ---------
                                       1/3
                                      C
```

We can now put our results together to obtain a solution to the problem. To multiply all of the numeric constants together, we use evalf.

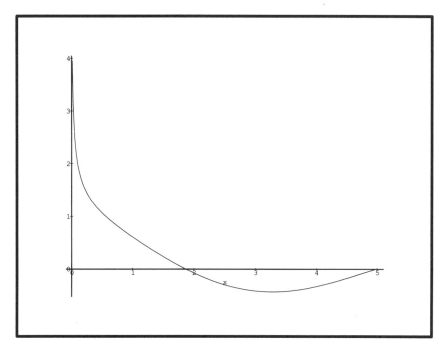

FIGURE 34. `plot(J(-1/3,x), x=0..5);`

`> evalf(");`

$$L = \frac{1.986352707}{C^{1/3}}$$

Finally, we replace C by the original set of symbolic parameters.

`> subs( C=W/(M*A), " );`

$$L = 1.986352707 \frac{M^{1/3} A^{1/3}}{W^{1/3}}$$

The result shows a simple relationship between the critical length of a rod and the elasticity of its material, its circular cross-section, and the weight of the rod per unit length. It would not have been as easy to derive this result without using a combination of symbolic math operations, numeric methods, and two-dimensional plotting.

6.5   Zeros of Bessel functions

We wish to compute a table of the first N positive zeros of the Bessel function of the first kind $J_\nu(x)$ for various integer orders ν.

Using standard notation (see *Handbook of Mathematical Functions* by Abramowitz and Stegun [AS65]), let $j_{\nu,s}$ denote the sth positive zero of $J_\nu(x)$.

The Maple name of the function is `BesselJ`. Use the `alias` facility to specify a shorter name.

```
> alias(J=BesselJ):
```

First let us look at a plot of $J_\nu(x)$ for $\nu = 0, 1, 2$ (Figure 35).

```
> plot({J(0,x), J(1,x), J(2,x)}, x = 0..10):
```

Consider the problem of locating $j_{\nu,1}$, the first positive zero of $J_\nu(x)$. We wish to determine an interval containing $j_{\nu,1}$ and no other zeros of $J_\nu(x)$ so that we can compute the appropriate root using `fsolve`. From the plot, we see that $j_{0,1}$ (the first zero of $J_0(x)$) lies in $[2, 3]$.

Abramowitz and Stegun give the following asymptotic formula for $j_{\nu,1}$ valid as $\nu \to \infty$:

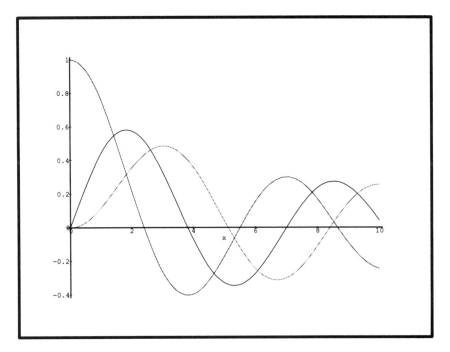

FIGURE 35. $J_\nu(x)$ for $\nu = 0, 1, 2$

$$\nu + 1.8557571\nu^{1/3} + 1.033150\nu^{-1/3} + O(\nu^{-1})$$

We can state that $j_{\nu,1}$ lies in the interval $[2, h(\nu)]$ where a computable formula for $h(\nu)$ can be obtained from the asymptotic formula. For example, if we define

$$h(\nu) = \nu + 1.9 * \nu^{1/3} + 1$$

then since this is an over-estimate of the asymptotic formula (for large ν) it will be true that $j[\nu,1]$ lies in $[2, h(\nu)]$ for ν large enough. Experimentation with small values of ν shows that this formula for $h(\nu)$ gives a valid interval $[2, h(\nu)]$ containing $j_{\nu,1}$ for all integers $\nu > 0$.

If we have computed a zero $j_{\nu,s-1}$ then the interval in which to search for the next zero $j_{\nu,s}$ is $[j_{\nu,s-1}, j_{\nu,s-1}+ incr]$ where we discover that a good value for $incr$ is 4.

The following procedure ZerosBesselJ computes the zeros $j_{\nu,s}$ for $\nu = 0, 1, ..., maxv$ and $s = 1, 2, ..., maxs$.

```
> ZerosBesselJ := proc (maxv, maxs)
>       local j, incr, v, h, s;
>       j := array(0..maxv, 1..maxs);
>       incr := 4.0;
>       for v from 0 to maxv do
>           h := evalf(v + 1.9*v^(1/3) + 1);
>           if v = 0 then
>               j[v,1] := fsolve(BesselJ(v,x), x, 2.0 .. 3.0)
>           else
>               j[v,1] := fsolve(BesselJ(v,x), x, 2.0 .. h)
>           fi;
>           for s from 2 to maxs do
>               j[v,s] := fsolve(BesselJ(v,x), x, j[v,s-1] .. j[v,s-1]+incr)
>           od
>       od;
>       RETURN( eval(j) )
> end:
```

Now we are ready to compute a table of the zeros of $J_\nu(x)$. Set the global variable Digits to whatever number of digits is desired.

```
> Digits := 10;
                                Digits := 10
```

```
> j := ZerosBesselJ(5,5):
```

Now $j_{\nu,s}$ is the entry of row ν, column s of the array. For printing purposes, convert to the Maple type matrix.

```
> convert(j, matrix);
    [ 2.404825558  5.520078110  8.653727913  11.79153444  14.93091771 ]
    [                                                                  ]
    [ 3.831705970  7.015586670  10.17346814  13.32369194  16.47063005 ]
    [                                                                  ]
    [ 5.135622302  8.417244140  11.61984117  14.79595178  17.95981950 ]
    [                                                                  ]
    [ 6.380161896  9.761023130  13.01520072  16.22346616  19.40941523 ]
    [                                                                  ]
    [ 7.588342435  11.06470949  14.37253667  17.61596605  20.82693296 ]
    [                                                                  ]
    [ 8.771483816  12.33860420  15.70017408  18.98013388  22.21779990 ]
```

6.6 Stock market analysis through linear algebra

A study of the London stock market, using the London Financial Times over a period of 1097 trading days, was found to fit the following transition matrix P

	I	D	N
I	586/1000	70/1000	79/1000
D	73/1000	639/1000	64/1000
N	341/1000	291/1000	857/1000

where I means an increase in stock value, D a decrease, and N no change. The numbers in the P matrix are probabilities. For example, the probability of going from no change to an increase is 79/1000. The problem is to find the probability that a stock will increase, decrease, or remain the same in the long term. This is the vector x of probabilities such that

$$Px = x.$$

The solutions x are the eigenvectors of the matrix P. We can compute them directly using the `eigenvects` command in the `linalg` package in the Maple library. However, we will compute the eigenvectors a step at a time to illustrate the calculations that need to be done. First let's enter the P matrix and compute its eigenvalues.

```
> with(linalg):
> P:=evalm(1/1000*matrix(3,3,[586,70,79,73,639,64,341,291,857]));
```

$$
P := \begin{bmatrix}
\dfrac{293}{500} & 7/100 & \dfrac{79}{1000} \\[2mm]
\dfrac{73}{1000} & \dfrac{639}{1000} & 8/125 \\[2mm]
\dfrac{341}{1000} & \dfrac{291}{1000} & \dfrac{857}{1000}
\end{bmatrix}
$$

```
> e := eigenvals(P);
```

$$
e := 1,\ \frac{541}{1000} + 1/200\ 43^{1/2},\ \frac{541}{1000} - 1/200\ 43^{1/2}
$$

For each eigenvalue λ, we want to solve $Px = \lambda x$, or equivalently solve

$$
(P - \lambda I)x = 0.
$$

Given a scalar λ, we can construct the matrix $P - \lambda I$, known as the *characteristic matrix* of P. A basis for the null space of the characteristic matrix of P is a set of vectors satisfying the above equation. Any linear combination of basis vectors will also satisfy the equation. We use Maple's `charmat` and `nullspace` functions to find a basis for the null space.

```
> v := nullspace( charmat(P,e[1]) );
```

$$
v := \{[\ 1,\ \frac{32263}{32999},\ \frac{144344}{32999}\ ]\}
$$

We see that there is one vector in the basis. Now doing this for the other two eigenvalues, and evaluating in floating point, we have

```
> nullspace(charmat(P,e[2]));
```

$$
\{[\ -\,1/5\ \frac{-3262 + 395\ 43^{1/2}}{-75 + 143\ 43^{1/2}},\ -1/5\ \frac{2887 + 320\ 43^{1/2}}{-75 + 143\ 43^{1/2}},\ 1\ ]\}
$$

```
> evalf(");
```

$$
\{[\ .1557438525,\ -1.155743853,\ 1.\ ]\}
$$

```
> nullspace(charmat(P,e[3]));
```

$$
\{[\ -\,1/5\ \frac{3262 + 395\ 43^{1/2}}{75 + 143\ 43^{1/2}},\ -1/5\ \frac{-2887 + 320\ 43^{1/2}}{75 + 143\ 43^{1/2}},\ 1\ ]\}
$$

```
> evalf(");
```

$$
\{[\ -1.155743852,\ .1557438524,\ 1.\ ]\}
$$

We see that only the eigenvector for the eigenvalue 1 has all real and positive entries. The other eigenvectors can be discarded since we cannot have negative probabilities. To convert the solution vector v to probabilities we normalize so that the sum of the values is 1.

```
> v := v[1];
```

$$\left[\ 1,\ \frac{32263}{32999},\ \frac{144344}{32999}\ \right]$$

```
> evalm(v/(v[1]+v[2]+v[3]));
```

$$\left[\ \frac{32999}{209606},\ \frac{32263}{209606},\ \frac{72172}{104803}\ \right]$$

Evaluating numerically to 3 digits precision we obtain the probabilities
```
> evalf(",3);
```

$$[\ .157,\ .154,\ .689\ ]$$

Thus, the probabilities are that, in the long run, a stock will increase 16% of the time, decrease 15% of the time, and stay the same 69% of the time. We conclude that the value of stocks is increasing in the long term!

6.7   Primitive trinomials

This example shows how to find primitive trinomials over the finite field with two elements. Primitive polynomials are used in the generation of error correcting codes. Trinomials are of special interest since they result in faster computation of these codes.

A primitive trinomial of degree n over the finite field with two elements is a primitive polynomial of the form

$$x^n + x^m + 1$$

whose coefficients are in the algebraic domain $GF(2)$.

For a polynomial $a(x)$ of degree k to be primitive with respect to $GF(2)$, it must be irreducible (cannot be decomposed into a product of polynomials of lesser degree whose coefficients are all from $GF(2)$). Furthermore, the sequence $x^i \bmod a(x)$ in $Z_2[x]$ for $i = 1, 2, \ldots 2^k - 1$ must contain no repetitions.

To find all primitive trinomials over $GF(2)$ with degrees between 10 and 20, we use nested for loops which conduct an exhaustive search, using Powmod and Irreduc functions from the Maple library.

```
> readlib(ifactors):
> for d from 10 to 20 do
>       n := ifactors(2^d-1);
>       trinomials := NULL;
>       for i from 1 to d-1 do
>           candidate := x^d+x^i+1;
>           if not Irreduc(candidate) mod 2 then next fi;
>           isPrimitive := true;
>           for j in n[2] while isPrimitive do
>               if Powmod(x,(2^d-1)/j[1],candidate,x) mod 2 = 1 then
>                   isPrimitive := false
>           fi;
>       od;
>       if isPrimitive then trinomials := trinomials, candidate fi;
```

```
>      od;
>      if trinomials <> NULL then print(degree .d, trinomials) fi;
> od:
```

$$\text{degree } 10, \ x^{10} + x^3 + 1, \ x^{10} + x^7 + 1$$

$$\text{degree } 11, \ x^{11} + x^2 + 1, \ x^{11} + x^9 + 1$$

$$\text{degree } 15, \ x^{15} + x + 1, \ x^{15} + x^4 + 1, \ x^{15} + x^7 + 1, \ x^{15} + x^8 + 1,$$

$$x^{15} + x^{11} + 1, \ x^{15} + x^{14} + 1$$

$$\text{degree } 17, \ x^{17} + x^3 + 1, \ x^{17} + x^5 + 1, \ x^{17} + x^6 + 1, \ x^{17} + x^{11} + 1,$$

$$x^{17} + x^{12} + 1, \ x^{17} + x^{14} + 1$$

$$\text{degree } 18, \ x^{18} + x^7 + 1, \ x^{18} + x^{11} + 1$$

$$\text{degree } 20, \ x^{20} + x^3 + 1, \ x^{20} + x^{17} + 1$$

The `ifactors` function returns the factorization of an integer n in the format $[s, [[p_1, e_1], \ldots, [p_m, e_m]]]$ such that $n = s \times p_1^{e_1} \times \ldots \times p_m^{e_m}$, where p_i are prime integers and the e_i is the number of times p_i divides n.

The `Irreduc` function tests if a polynomial is irreducible mod p. In `Powmod`(a, n, b, x) mod p, the `Powmod` function computes the remainder of a^n divided by b mod p for large n. It does this efficiently through the use of binary powering.

For higher degree, the factorization of $2^d - 1$ will become the bottleneck. We can improve the speed of this particular integer factorization by considering first the factorization of the polynomial

$$x^d - 1$$

For example, for $d = 50$ we have

```
> factor(x^50-1);
```

$$(x - 1) (x^4 + x^3 + x^2 + x + 1) (x^{20} + x^{15} + x^{10} + x^5 + 1) (x + 1)$$

$$(1 - x + x^2 - x^3 + x^4) (1 - x^5 + x^{10} - x^{15} + x^{20})$$

Therefore the factors of the integer $2^{50} - 1 = 1125899906842623$ are the factors of the integers

```
> subs(x=2,[op(")]);
                    [1, 31, 1082401, 3, 11, 1016801]
```

which factor as follows

```
> map(ifactor,");
            [1, (31), (601) (1801), (3), (11), (251) (4051)]
```

```
> convert(",*);
            (31) (601) (1801) (3) (11) (251) (4051)
```

Using this idea we can factor a number of the form $2^n - 1$ faster via first factoring the polynomial $x^n - 1$. The following Maple procedure factors the integer $2^n - 1$ efficiently:

```
> fastifactors := proc(n) local f;
>      f := factor(x^n-1);
>      f := subs(x=2,[op(f)]);
>      f := map(ifactor,f);
>      ifactors(convert(f,*));
> end:
```

A second improvement that will save half the work comes from the fact that if $x^n + x^m + 1$ is primitive then its reciprocal $x^n + x^{(n-m)} + 1$ is also primitive.

Including both ideas we have the following program to compute all primitive trinomials of degree 90 to 100.

```
> for d from 90 to 100 do
>      n := fastifactors(d);
>      trinomials := NULL;
>      for i from 1 to d/2 do
>         candidate := x^d+x^i+1;
>         if not Irreduc(candidate) mod 2 then next fi;
>         isPrimitive := true;
>         for j in n[2] while isPrimitive do
```

```
>          if Powmod(x,(2^d-1)/j[1],candidate,x) mod 2 = 1 then
>              isPrimitive := false
>          fi;
>       od;
>       if isPrimitive then
>           trinomials := trinomials, candidate;
>           if i<d/2 then trinomials := trinomials, x^d+x^(d-i)+1 fi
>       fi;
>    od;
>    if trinomials <> NULL then print(degree .d, trinomials) fi;
> od:
```

$$\text{degree 93, } x^{93} + x^{2} + 1, \ x^{93} + x^{91} + 1$$

$$\text{degree 94, } x^{94} + x^{21} + 1, \ x^{94} + x^{73} + 1$$

$$\text{degree 95, } x^{95} + x^{11} + 1, \ x^{95} + x^{84} + 1, \ x^{95} + x^{17} + 1, \ x^{95} + x^{78} + 1$$

$$\text{degree 97, } x^{97} + x^{6} + 1, \ x^{97} + x^{91} + 1, \ x^{97} + x^{12} + 1, \ x^{97} + x^{85} + 1,$$

$$x^{97} + x^{33} + 1, \ x^{97} + x^{64} + 1, \ x^{97} + x^{34} + 1, \ x^{97} + x^{63} + 1$$

$$\text{degree 98, } x^{98} + x^{11} + 1, \ x^{98} + x^{87} + 1, \ x^{98} + x^{27} + 1, \ x^{98} + x^{71} + 1$$

$$\text{degree 100, } x^{100} + x^{37} + 1, \ x^{100} + x^{63} + 1$$

6.8   Computations on the $3n+1$ conjecture

Consider the sequence defined recursively by

$$n_{i+1} = \begin{cases} 3n_i + 1 & \text{if } n_i \equiv 1 \pmod 2 \\ n_i/2 & \text{if } n_i \equiv 0 \pmod 2 \end{cases}$$

For example, if $n_0 = 7$ the sequence starts as $7, 22, 11, 34, 17, 52, 26, 13, 40, 20, 10, 5, 16, 8, 4, 2, 1, 4, 2, 1, 4, \ldots$. It is obvious that this sequence will cycle through the values $4, 2, 1$ forever. The $3n+1$ conjecture states that for *any* starting positive integer n_0, the sequence eventually reaches 1 and repeats as above.

Computations on the $3n+1$ sequence have probably consumed more CPU time than any other number theoretic conjecture. The statement of the problem is so simple, one feels compelled to write programs to test it. This sequence, attributed to Lothar Collatz, has been given various names, including Ulam's conjecture, Syracuse's problem, Kakutani's problem and Hasse's algorithm.

Jeffrey Lagarias published an excellent article summarizing most of the work done on this problem [Lag85].

In this example we show how to efficiently compute iterations of this sequence using Maple. The programs that we will develop can compute thousands of iterations over terms which are thousands of digits long, reasonably quickly.

We provide two main functions for working with these sequences: `DistTo1` computes the number of iterations that it will take for its argument to reach the value 1 for the first time, and `Iterate` iterates its first argument the number of times specified in its second argument. Both functions use a recursive modular computation of the terms which requires some additional explanation.

Let $\sigma(n)$ denote the number of iterations required to reduce n to 1, or what we call the distance to 1. The following relations are known or easy to derive:

$$\sigma(n2^k) = \sigma(n) + k$$

$$\sigma(n2^k - 1) = \sigma(n3^k - 1) + 2k$$

$$\sigma(n4^k + 1) = \sigma(n3^k + 1) + 3k$$

It is easy to see that for any value of n, the first k iterations, which are divisions by 2, depend exclusively on the value of $n \pmod{2^k}$. For example,

$$\sigma(8n + 3) = \sigma(9n + 4) + 5$$

$$\sigma(8n + 5) = \sigma(3n + 2) + 4$$

It should also be observed that the coefficient of n in the resulting term is always a power of 3. Furthermore, this power of 3 corresponds exactly to the number of $3n + 1$ iterations necessary while doing the first k iterations (divisions by 2).

This suggests that we could run the iteration in symbolic terms, for an arbitrary value of k by computing the coefficients for the resulting iterations. This strategy has two immediate advantages. First, for very large numbers, we will need to do the computation with long numbers only once for every k division by 2 steps. Secondly, we could easily tabulate all the possible values for small values of k, and hence save in the computation of the coefficients themselves. In mathematical terms, what we will compute are the functions a_k and b_k defined by the following equation:

$$\sigma(2^k n_1 + n_0) = \sigma(3^{b_k(n_0)} n_1 + a_k(n_0)) + k + b_k(n_0)$$

By analyzing two steps of such an iteration we can derive the function a_{2k} and b_{2k} in terms of a_k and b_k. Let

$$n = n_1 2^k + n_0, q = n_1 3^{b_k(n_0)} + a_k(n_0).$$

Then

$$a_{2k}(n) = \lfloor q/2^k \rfloor 3^{b_k(q \bmod 2^k)} + a_k(q \bmod 2^k), \quad b_{2k}(n) = b_k(n_0) + b_k(q \bmod 2^k).$$

As it turns out, this gives us a very efficient method for computing several iterations of the recurrence at once. The main program to compute the $\sigma(n)$ function will compute k as an appropriate power of 2, as large as possible, but guaranteeing that the series will not reach 1 in less than k steps. This is achieved simply by insuring that $2^k < n$. The function definition, argument checking and initialization look like:

```
> DistTo1 := proc(n) local its, t, qt, u, rt, bits;
```

```
>         if not type([args],[integer]) or n < 1 then
>             ERROR(`invalid arguments`) fi;
>         its := 0;
>         t := n;
>         do
```

If the argument is small enough (or becomes small enough), compute the result directly.
```
>             if t < 10 then RETURN( op(t,[0,1,7,2,5,8,16,3,19,6]) + its ) fi;
```

Now compute $2^{bits} < t$ by comparing the lengths of their decimal representations (the iteration is stopped before exceeding t so that we save one power of 2 computation, the largest).
```
>             bits := 3;
>             while 2*length(pow2(bits)) < length(t) do bits := 2*bits od;
```

The function iquo returns the quotient and stores the remainder in its third argument.
```
>             qt := iquo(t,pow2(bits),'rt');
>             u := CollatzMod(rt,bits);
```

The values a_{bits} and b_{bits} are in the list u; now add the number of iterations and compute the next value of t.
```
>             its := its+u[2]+bits;
>             t := qt*3^u[2]+u[1];
>         od;
> end:
```

The function which computes a_{bits} and b_{bits} is written recursively. The bottom of the recursion is entered as fixed constants at the end of the definition. The body of this function just follows the above mathematical definition.
```
> CollatzMod := proc(r,b)
> local b2, q1, q2, u1, u2, res;
>     b2 := iquo(b,2);
```

The function irem returns the remainder and stores the quotient in its third argument.
```
>     u1 := CollatzMod(irem(r,pow2(b2),'q1'),b2);
>     u2 := CollatzMod(irem(q1*3^u1[2]+u1[1],pow2(b2),'q2'),b2);
>     res := [q2*3^u2[2]+u2[1], u1[2]+u2[2]];
```

If b is less or equal to 12, then we will remember any computed value for the future. This will effectively create a dynamic table (which will only grow as needed) with at most $4175 = 2^{12} + 2^6 + 2^3 + 2^2 + 2^1$ entries.
```
>     if b <= 12 then CollatzMod(args) := res fi;
>     res
> end:
```

Now we will define all the possible cases for 3 or less bits. This will terminate the recursion of the CollatzMod function. (Notice that b is always of the form 3×2^k).
```
> CollatzMod(0,1) := [0,0]:  CollatzMod(1,1) := [2,1]:
> CollatzMod(0,2) := [0,0]:  CollatzMod(1,2) := [1,1]:
> CollatzMod(2,2) := [2,1]:  CollatzMod(3,2) := [8,2]:
> CollatzMod(0,3) := [0,0]:  CollatzMod(1,3) := [2,2]:
> CollatzMod(2,3) := [1,1]:  CollatzMod(3,3) := [4,2]:
> CollatzMod(4,3) := [2,1]:  CollatzMod(5,3) := [2,1]:
> CollatzMod(6,3) := [8,2]:  CollatzMod(7,3) := [26,3]:
```

The above functions will use some powers of 2 repeatedly. Since these powers of 2 are rare (just the ones of the form $2^{3 \times 2^b}$), it is worthwhile computing them efficiently and remembering the past values.
```
> pow2 := proc(b) option remember;
```

```
> if b <= 24 then 2^b else pow2(b/2)^2 fi end:
```

To test Dist1, compute some values and try to verify some of the above identities.

```
> DistTo1( 2^200 );
```

$$200$$

```
> DistTo1( 2^100-1 ) = DistTo1( 3^100-1 ) + 200;
```

$$1465 = 1465$$

```
> DistTo1( 25*4^100+1 ) = DistTo1( 25*3^100+1 ) + 300;
```

$$1562 = 1562$$

The function Iterate will start with a value x , perform n Collatz iterations, and return the resulting term. It is written using the same ideas of modular computation as above, plus a provision for terminating quickly once the iteration reaches its cycle at 4,2,1.

```
> Iterate := proc( x, n )
> local bits, qt, rn, rt, rx, u;
>      if not type([args],[posint,posint]) then
>          ERROR(`invalid arguments`) fi;
>      rn := n;  rx := x;
>      while rn > 0 do
>          if rx=1 then RETURN( op( modp(rn,3)+1, [1,4,2] ) )
>          elif rx=2 then RETURN( op( modp(rn,3)+1, [2,1,4] ) )
>          elif rx=4 then RETURN( op( modp(rn,3)+1, [4,2,1] ) )
>          elif rn < 6 then
>              if irem(rx,2,'qt')=1 then rx := 3*rx+1 else rx := qt fi;
>              rn := rn-1
>          else bits := 3;
>              while 2*length(pow2(bits)) < length(t) and 4*bits < rn
>                  do bits := 2*bits od;
>              qt := iquo(rx,pow2(bits),'rt');
>              u := CollatzMod(rt,bits);
>              rn := rn-u[2]-bits;
>              rx := qt*3^u[2]+u[1];
>          fi
>      od;
>      rx
> end:
```

Now we can try some examples of this function:

```
> Iterate(12345678901234567890,100);
```

$$2790345133793393$$

```
> Iterate( 2^100-1, 2^100 );
```

$$1$$

The next expression should return 1 for any value of x.

```
> x := 70524774652746589746574456924727562;
                    x := 70524774652746589746574456924727562
> Iterate( x, DistTo1(x) );
```

$$1$$

The following timing information was generated on a DEC 5000 workstation. Notice that $3^{50000} - 1$ is an integer 23,857 digits long.

```
> st := time():  DistTo1( 3^50000-1 );  time()-st;
```

567858
518.350

Finally, let's plot the length of the sequences starting with n for n up to 4000 (Figure 36).
```
> plot( [seq( op( [i,DistTo1(i)]), i=1..4000)], style=POINT);
```

6.9   A numerical approximation problem

We wish to develop a procedure for the efficient numerical evaluation, to seven digits accuracy, of the following "Gamma integral" function:

$$Gi(x) = \int_0^x 1/\Gamma(t)\, dt, \qquad \text{for } 0 \le x \le 4$$

where $\Gamma(t)$ is the GAMMA function known to Maple. We can specify Gi in Maple using the arrow operator notation.

```
> Gi := x -> int(1/GAMMA(t), t=0..x);
```

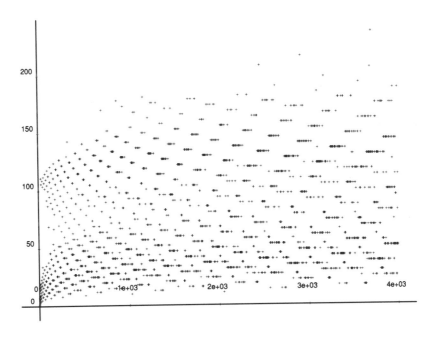

FIGURE 36. Sequence lengths for n up to 4000.

```
                              x
                             /
                            |      1
     Gi := x ->             |   -------- dt
                            |   GAMMA(t)
                            /
                            0
```

We can evaluate $Gi(x)$ at any real value x through the numerical integration built into `evalf`. For example:

```
> evalf(Gi(1.2));
                          .7510244314
```

The procedure which we will develop will be able to evaluate this function more efficiently than `evalf`'s numerical integration approach.

6.9.1   Transforming to a function $f(x)$ which is nonzero on $[0, 4]$

From the definition we can see that $Gi(0) = 0$. Furthermore, the series expansion of $Gi(x)$ about $x = 0$ has the form

```
> series(Gi(x), x=0, 5);
              2                3               2            2   4        5
       1/2 x   + 1/3 gamma x   + (- 1/48 Pi  + 1/8 gamma ) x   + O(x )
```

Let us divide by x^2 and work with the function

```
> f := x -> Gi(x)/x^2;
                              Gi(x)
               f := x -> -----
                                2
                               x
```

Let's look at a plot of f (see Figure 37).

```
> plot(f, 0..4);
```

f has a series of the form

```
> series(f(x), x=0, 5);
                              2               2   2        3
       1/2 + 1/3 gamma x + (- 1/48 Pi  + 1/8 gamma ) x   + O(x )
```

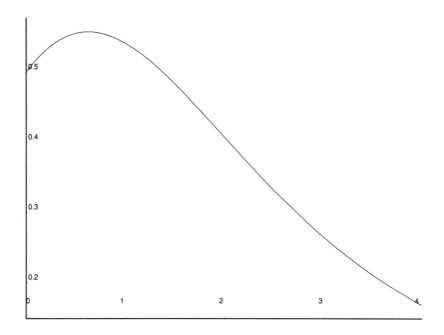

FIGURE 37. The function $f(x)$.

6.9.2   Approximations derived from Taylor series

Efficient evaluation procedures are usually based on polynomial approximations, or rational function approximations (i.e. quotients of polynomials). The reason is that a polynomial or a rational function can be evaluated directly using only the four basic arithmetic operations: addition, subtraction, multiplication, and division.

As our first approximation to the function $f(x)$, we compute the Taylor polynomial of degree 8 about the midpoint.

```
> s := map(evalf, series(f(x), x=2, 9));
                                                        2
s := .4065945998 - .1565945998 (x - 2) + .00209790791 (x - 2)
                  3                        4                         5
    + .01762626393 (x - 2)  - .006207547124 (x - 2)  + .00057335661 (x - 2)
                     6                       7                          8
    + .00024331164 (x - 2)  - .000100105329 (x - 2)  + .0000141421155 (x - 2)
                 9
    + O((x - 2) )
> TaylorApprox := convert(s, polynom):
```

The command convert(s, polynom) forms a polynomial by dropping the order term.

For convenience, we convert the approximation to functional form to correspond with the form specified for f.

```
> TaylorApprox := unapply(TaylorApprox, x);
```

$$
\begin{aligned}
\text{TaylorApprox} := x \to\ & .7197837994 - .1565945998\ x + .00209790791\ (x - 2)^2 \\
& + .01762626393\ (x - 2)^3 - .006207547124\ (x - 2)^4 + .00057335661\ (x - 2)^5 \\
& + .00024331164\ (x - 2)^6 - .000100105329\ (x - 2)^7 + .0000141421155\ (x - 2)^8
\end{aligned}
$$

For our next approximation to $f(x)$, we compute the Pade rational approximation of degree $(4,4)$.

```
> PadeApprox := convert(s, ratpoly):
> PadeApprox := unapply(PadeApprox, x);
```

$$
\begin{aligned}
\text{PadeApprox} := x \to\ & (.3410347493 + .03277992524\ x - .006127825127\ (x - 2)^2 \\
& + .004529909540\ (x - 2)^3 - .0004315062579\ (x - 2)^4)\ \big/\ (.0684847996 \\
& + .4657576002\ x + .1591496591\ (x - 2)^2 + .02668138298\ (x - 2)^3 \\
& + .003469681202\ (x - 2)^4)
\end{aligned}
$$

By definition, the Taylor series expansion of PadeApprox about x = 2 agrees with the Taylor series s up to the degree 8 term.

If we plotted $f(x)$, the degree 8 Taylor polynomial, and the $(4,4)$ Pade approximation on the same graph, we would find that the three curves are almost indistinguishable from one another.

Let's look at the error curve for the Pade approximation. Because our definition of f does not work at $x = 0$, we use the limit of f at that point for the definition of $f(0)$.

```
> limit(f(x), x=0);
```
$$1/2$$
```
> PadeError := proc(x)
>     if x = 0 then 0.5 - PadeApprox(0)
>     else evalf(f(x)) - PadeApprox(x)
>     fi
> end:
```

```
> plot(PadeError, 0..4);
```

The plot of PadeError appears in Figure 38.

We see that the error is largest at the left end-point of the interval. The size of the error at $x = 0$ is

```
> PadeError(0);
```
$$.0003537682$$

This error is not small enough to satisfy our goal of obtaining approximations which are accurate to seven digits.

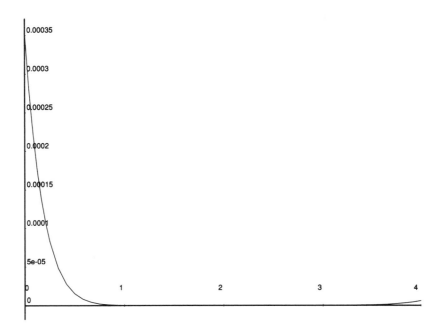

FIGURE 38. Error curve for $(4, 4)$ Pade approximation.

6.9.3   Approximations derived from Chebyshev series

In general, better approximations on an interval can be obtained using Chebyshev series expansions rather than Taylor series expansions. The Chebyshev series expansion of $f(x)$ on $[0, 4]$ can be computed using Maple's built-in `chebyshev` function. The references to T in the result are to the Chebyshev polynomials, following the naming conventions of Maple's `orthopoly` package.

```
> fproc := proc(x) if x=0 then 0.5 else evalf(f(x)) fi end:

> Cheb := chebyshev(fproc, x=0..4, Float(1,-6));
  Cheb := .3792062743 T(0, 1/2 x - 1) - .2026328140 T(1, 1/2 x - 1)
       - .03690648364 T(2, 1/2 x - 1) + .03701314315 T(3, 1/2 x - 1)
       - .008889441430 T(4, 1/2 x - 1) - .0001497893366 T(5, 1/2 x - 1)
       + .0006429746208 T(6, 1/2 x - 1) - .0001706779494 T(7, 1/2 x - 1)
                                                           -5
       + .00001269172839 T(8, 1/2 x - 1) + .4398749287*10   T(9, 1/2 x - 1)
                         -5                                 -6
       - .1562841398*10    T(10, 1/2 x - 1) + .2049805401*10    T(11, 1/2 x - 1)
```

In the Chebyshev series computation, we have truncated the series by dropping all terms with coefficients less than 10^{-6}. We discover that this yields a degree 11 approximation on $[0, 4]$.

Let's look at the error curve for the Chebyshev series approximation.

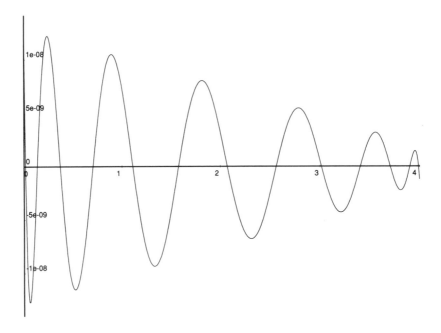

FIGURE 39. Error curve for Chebyshev series.

```
> Cheb := unapply(Cheb, x):
> with(orthopoly,T):
> ChebError := proc(x)
>      if x = 0 then 0.5 - Cheb(0)
>      else evalf(f(x)) - Cheb(x)
>      fi
> end:
> plot(ChebError, 0..4);
```

The plot of ChebError appears in Figure 39. Oscillating error curves are typical of Chebyshev series approximations.

Just as Pade rational approximations are generally better than truncated Taylor series, so too we can compute rational approximations called Chebyshev-Pade approximations which are generally better than truncated Chebyshev series for approximating a function on an interval. The reader is referred to the on-line help ?convert[ratpoly] for further information on computing such approximations.

We find that the error is largest at the left end-point of the interval, with the value of the maximum error being

```
> ChebError(0);
```

$$.127*10^{-7}$$

Therefore, we have determined an efficient evaluation procedure for $f(x)$ which is accurate to within a maximum absolute error of 10^{-7}.

An evaluation procedure for efficiently computing values of the original function $Gi(x) = x^2 f(x)$ can be based on evaluating the simple function `x^2 * Cheb(x)`. For even greater efficiency, such an expression could be evaluated using `evalhf`, or translated to a language such as Fortran or C using the facilities described in Section 2.18.

This example has shown how the facilities in Maple can be used very effectively to develop procedures for the numerical evaluation of functions. Specifically, the ability to generate series expansions, to compute rational function approximations, and to convert to various special forms, combined with the powerful interactive tool of viewing plots of the approximations and their error curves, provides an ideal environment for such tasks.

6.10   Reading more about Maple problem-solving techniques

To read more about problems solved by Maple, see *The Maple Technical Newsletter*[Sco92], *The Roots Newsletter*[Sof92], and the list of articles in Appendix B.

Chapter 7
Global access to Maple information

Where can you turn when on-line `help` (?), this tutorial, or the *Maple V Language Reference Manual* do not help you overcome difficulties in using Maple? Who can you talk to about solving your special kinds of problems with Maple? In this section, we discuss additional sources of information available in person, in print, and through electronic media.

7.1 New users' problems

Beginner's and installation instructions typically come in the Maple documentation given to you, either by the distributor or system administrator, or in the package containing the software. Such documentation, as well as locally available expertise, is the best place to look for answers to problems having to do with your specific computer system. If you have purchased a copy of Maple directly from a distributor, consider contacting their technical support staff for help. Examples of such problems might be:

1. The computer is on and will do other tasks for you, but you can't start up Maple.

2. You start up Maple, but it does not respond to commands you give it.

3. Many "funny characters" appear on the display.

4. Plots aren't drawn correctly, or you can't figure out how to get them printed.

5. On-line help does not work properly.

6. Built-in commands do not work.

Even in this era of "user-friendly" computer applications you can encounter difficulty when doing sophisticated computing such as that available through Maple. As with learning any new skill, you should be prepared to experiment and to practice, and to get feedback by mulling over your experiences, and talking with others. Be prepared to spend time experimenting with the computer, and seek out other Maple users who could be your "coach" or compatriot. Approaching the process of learning how to use new computer software with patience and good humor is essential!

7.2 The community of Maple users

Maple users in your area can be a valuable source of information. They may be in your office building, laboratory, or class; or easily contacted on the phone or through local computer users' groups. You can also be part of the world-wide community of Maple users through electronic mail and computer network news services. For example, if you have access to the Internet, sending an electronic mail message to maple_group@daisy.waterloo.edu will reach a world-wide readership of Maple users, fans, and experts. Someone out there may be able to help you.   You can subscribe to this on-going group discussion of Maple problems and solutions by sending electronic mail to maple_group@daisy.waterloo.edu with a subscription request.

Another electronic forum that addresses symbolic computation (both in Maple and by other systems) is sci.math.symbolic carried by Usenet. To subscribe to this group, you need to have access to a computer facility that gets a Usenet "news feed". Many colleges, universities, research laboratories, government facilities and corporations do so.

7.3   What to do when the answer seems wrong

Maple undergoes extensive testing before each release. Test programs are run for each Maple command. Maple also undergoes pre-release testing by volunteers from the user community. However, such testing does not uncover all mistakes. You may discover what appears to be a mathematically incorrect answer, or a result that contradicts the documentation. Here are some steps to follow when you encounter something suspicious:

1. Convince yourself that Maple is computing an incorrect answer. Often times this is easier to do than figuring out what the right answer on your own. For example, if you are solving an algebraic equation, you can see whether the alleged solution works by substituting it (using subs) into the equation. Checking the validity of a solution is usually so inexpensive and easy to do that you should consider doing it as standard practice for most computations.

2. If the computed result does not meet your expectations, recheck the documentation for the built-in Maple procedures you are using. You may discover that a procedure works differently than the way you thought it did. One way to know that a built-in procedure is defective is if you can get it to perform correctly on some other problem, such as one copied from the on-line help documentation or the *Maple V Library Reference Manual*. If you cannot get a built-in procedure to work at all, it is indicative of an installation problem. Refer to your Maple installation instructions or local computing consultants for further assistance.

3. Check values of global variables, especially if you are using some in your own Maple programming. Use mint (described in Section 3.16), if it is available, to determine if the programs you wrote are using global variables. Unintended settings of global variables may have influenced the computation of your result.

4. See if you can reproduce the anomaly by reading in a file with a short sequence of commands into a fresh Maple session. This will make it easy to demonstrate the mistake to others, and make clear what variables, values, procedures, and prior steps that the error depends upon.

5. Talk to another Maple user. This can often provide more insight into the nature of the problem, and provides a quick and easy "reality check".

6. Report the error to the parties responsible for your version of Maple, or send electronic mail to the Internet addresses support@maplesoft.on.ca or support@daisy.waterloo.edu.

7.4   Electronic access to user-contributed Maple software

If you can send and receive mail to Internet addresses (such as to `maple_group@daisy.waterloo.edu` discussed in Section 7.2), then you probably should be able to obtain free copies of Maple programs contributed by other Maple users, as well as fixes or extensions to versions of the Maple library released by the developers of the system. Such programs are available through two channels: Internet file transfer via the *ftp* command, and through electronic mail.

7.4.1   Information and programs via "netlib"

The CAN (Computer Algebra Netherlands) Expertise Center maintains a copy of Maple programs contributed by users, library fixes and updates, and other files of interest.

You can get a copy of any program or file by sending an electronic mail message to a specific Internet address: `maple-info@can.nl`. The first message you send should consist of a single line:

```
send info
```

Mail sent to this address is monitored and responses elicited automatically. In response to a `send info` message, you should get a reply describing how to request Maple files.   Example 119 shows a typical response to the `send info` request.

Example 119
Response from *netlib* to `send info` request

```
From: daemon@can.nl
Message-Id: <9111261944.AA05731@canb.can.nl>
To: bchar@king.mcs.drexel.edu
Subject: send info
Date: Tue, 26 Nov 91 20:44:46 +0100


              ===== GENERAL MAPLE NETLIB INDEX =====

Welcome to the MAPLE version of netlib, a system for the distribution of
software by electronic mail.  This index is the reply you'll get to the
single line message
        send index
or equivalently, the messages "help" or "info", sent to the network address
maple-netlib@can.nl .  The single line message can either be the subject of
the message or the body.

The available information is divided into a number of libraries, the names
of which are listed below.  You can obtain the full index to each library
and the individual members of each library as follows:
```

To examine the full index for any library send a request of the form:
          send index for patches5.0
To retrieve the file "latex" from the library patches5.0:
          send latex from patches5.0
To retrieve just the sizes of file names rather than the contents:
          send list of latex from patches5.0
To retrieve a complete list of the indices for all the libraries:
          send complete index.

When the file name is not used in more than one library, you can leave out
the library name. So to get the CAN information file in the library info,
it is sufficient to send the message
                send can

You may include several requests in a single piece of mail, but put each
on a separate line.

Send the requests to "maple-netlib@can.nl". You´ll be talking to a program,
so don´t expect it to understand much English. If you need to talk
to a human, or have problems with netlib, send mail to "can@can.nl".

Entries in the various indices are dated to indicate the date of the last
change. In particular, the dates in this index are changed
whenever any file in the relevant library is changed. So by periodically
perusing this index, you can determine if any of your files need updating,
or new material is available.

Information about changes and additions to the network library is sent at
regular intervals to participants in the Maple User Group. If you would like
to join this, please send a message requesting inclusion to
maple@daisy.waterloo.edu

For background information about netlib, see Jack J. Dongarra and Eric
Grosse, Distribution of Mathematical Software Via Electronic Mail, Comm.
ACM (1987) 30, 403-407.

                    -------Quick Summary of Contents---------
              --- The date given is the date of the last update ---

info   - 24 Nov 91 - Various information files (e.g., the CAN
                              information file)
patches4.3   - 23 Jul 91 - Patches correcting bugs in Maple version 4.3
patches5.0   - 24 Nov 91 - Patches correcting bugs in Maple version 5.0
share4.3   - 24 Nov 91 - Various pieces of code for Maple version 4.3
share5.0   - 24 Nov 91 - Various pieces of code for Maple version 5.0
tex   - 24 Nov 91 - Style files for printing TeX and LaTeX documents

After you receive an electronic mail message containing a Maple file, you should use the commands of your mail-reading program to save the message in a file. Using a text editor, you can then delete any superfluous information added to the file during its transformation into and transit as a mail message. Once you have deleted any "mail header" lines, you can read the file into a Maple session just as you would a file containing Maple procedures that you had created yourself.[1]

7.4.2   Access to Maple programs via *ftp*

Many workstations and larger computers in academia, research laboratories, and corporate or government offices can interact with each other through an amalgam of communication networks collectively referred to as the Internet. Many such computer systems have a program called *ftp*, short for *file transfer protocol* program. If you can run the *ftp* command on your computer, you can copy files of Maple software by doing the following:

1. First, open a connection to a computer containing the Maple files. The following addresses were valid as of the time when *First Leaves* was written:

Name	Alternative numerical address	Geographical location
daisy.waterloo.edu	129.97.140.58	Waterloo, Ontario Canada
neptune.inf.ethz.ch	129.132.101.33	Zurich, Switzerland

   On many computers, you can open the connection by the command ftp daisy.waterloo.edu (or ftp neptune.inf.ethz.ch). Other systems employ dialog boxes or menus to specify the computer's address. Consult the *ftp* documentation for your computer for further details.

2. Once the connection is established, log onto daisy (neptune), giving a user name, and a password. Connecting to the computer closest to you should provide the best *ftp* service. For the purposes of Maple file access, you may use anonymous as the user name, and your Internet mail address as the password (e.g. jpicard@ncc1701d.sf.gov).

3. After you have successfully logged onto the remote computer, you can use the commands ls and cd to list and change directories, respectively. The get *filename* command of *ftp* causes the specified file to be copied to your computer. The quit command closes the remote connection and concludes the *ftp* session.

Example 120

A ftp session: connection, log on, listing and changing directories, file copying, and sign-off

```
%ftp daisy.waterloo.edu
Connected to daisy.waterloo.edu.
220 daisy FTP server (Version 5.103 Sat Nov 17 21:57:34 EST 1990) ready.
Name (daisy.waterloo.edu:bchar): anonymous
331 Guest login ok, send ident as password.
Password:
230 Guest login ok, access restrictions apply.
```

---

[1]As this book was going to press, plans were being made by CAN to add "xnetlib" as an additional service to Maple files.

```
ftp> ls
200 PORT command successful.
150 Opening ASCII mode data connection for file list.
etc
bin
pub
maple
226 Transfer complete.
18 bytes received in 5e-06 seconds (3.5e+03 Kbytes/s)
ftp> cd maple
250 CWD command successful.
ftp> ls
200 PORT command successful.
150 Opening ASCII mode data connection for file list.
4.3
5.0
guide
help
README
address
mapletnsr
fixes
usage
tex
226 Transfer complete.
60 bytes received in 0.01 seconds (5.9 Kbytes/s)
ftp> get README
200 PORT command successful.
150 Opening ASCII mode data connection for README (2559 bytes).
226 Transfer complete.
local: README remote: README
2620 bytes received in 3.4 seconds (0.74 Kbytes/s)
ftp> quit
221 Goodbye.
%
```

Example 120 illustrates the transfer of a file named README from the maple subdirectory of those files accessible by "anonymous ftp" to daisy. If you have access to both *ftp* and *netlib*, you will probably prefer the former, since file transfers are usually faster via *ftp* than electronic mail.

7.5 Maple publications

Each issue of the *Maple Technical Newsletter* contains articles written by Maple users and the system's designers, describing Maple problem-solving techniques and other topics of general interest to Maple users. To obtain subscription information, write to:

Waterloo Maple Software
160 Columbia Street West
Waterloo, Ontario N2L 3L3
Canada

phone (519) 747-2373, fax (519) 747-5284
`info@maplesoft.on.ca` or `wmsi@daisy.waterloo.edu`

There are a growing number of books and articles addressing aspects of use of Maple and systems like it. Appendix B describes some of them available or announced at the time of publication.

Conclusion

Now that you have worked through *First Leaves*, you should be well-equipped for your mathematical problem-solving endeavors. While brevity has permitted us to cover only only key concepts and major features, you should feel confident that you can, like most experienced users of Maple, use the *Maple V Language Reference Manual* , on-line help (?), and experimentation with the program itself, to learn about further Maple features as your needs and ambitions grow.

We would like to hear your comments on Maple and on this tutorial. Please address correspondence to:

Symbolic Computation Group
Department of Computer Science
University of Waterloo
Waterloo, Ontario
Canada N2L 3G1

For those with access to electronic mail, our electronic address is:

maple@daisy.waterloo.edu

Appendix A

Bibliography

[AS65]   Milton Abramowitz and Irene A. Stegun, editors. *Handbook of Mathematical Functions with Formulas, Graphs, and Mathematical Tables*. Dover Publications, New York, 1965.

[BM77]   Garrett Birkhoff and Saunders MacLane. *A Survey of Modern Algebra*. Macmillan Co., 4 edition, 1977.

[CGG⁺91a] B.W. Char, K.O. Geddes, G.H. Gonnet, B.L. Leong, M.B. Monagan, and S.W. Watt. *Maple V Language Reference Manual*. Springer-Verlag, New York, 1991.

[CGG⁺91b] B.W. Char, K.O. Geddes, G.H. Gonnet, B.L. Leong, M.B. Monagan, and S.W. Watt. *Maple V Library Reference Manual*. Springer-Verlag, New York, 1991.

[CMS88]   B. Char, A. Macnaughton, and P. Strooper. Discovering Inequality Conditions in the Analytical Solution of Optimization Problems. In P. Gianni, editor, *International Symposium ISSAC '88*, New York, July 1988. Springer Verlag.

[Fat89]   R. Fateman. Lookup Tables, Recurrences and Complexity. In G. H. Gonnet, editor, *Proceedings of the ACM-SIGSAM 1989 International Symposium on Symbolic and Algebraic Computation*, New York, 1989. ACM Press.

[Knu71]   D. E. Knuth. An empirical study of Fortran programs. *Software – Practice and Experience*, 1:105–133, 1971.

[Kov86]   J. Kovacic. An algorithm for solving second order linear homogeneous differential equations. *Journal of Symbolic Computation*, 2(1), 1986.

[Lag85]   Jeffrey Lagarias. The 3x+1 Problem and Its Generalizations. *American Mathematical Monthly*, 92(1):3–23, January 1985.

[O'N91]   Peter V. O'Neil. *Advanced Engineering Mathematics*. Wadsworth Publishing, Belmont, California, third edition, 1991.

[Sco92]   T. Scott, editor. *The Maple Technical Newsletter*. Birkhauser, Boston, 1992. Issues 1 to 6 were published by Waterloo Maple Software.

[Sof92]   Waterloo Maple Software. *The Maple Roots Report*. Waterloo Maple Software, Waterloo, Ontario, 1990–1992.

Appendix B

Books and articles for Maple users

B.1 Some books for Maple users

[1] John W. Auer. *Maple Solutions Manual for Linear Algebra with Applications*. Prentice-Hall Canada, Scarborough, Ontario, 1991.

[2] William Bauldry and Joseph Fiedler. *Calculus Laboratories with Maple: A Tool, Not an Oracle*. Brooks/Cole Publishing, Pacific Grove, California, 1991.

[3] B.W. Char, K.O. Geddes, G.H. Gonnet, B.L. Leong, M.B. Monagan, and S.W. Watt. *Maple V Language Reference Manual*. Springer-Verlag, New York, 1991.

[4] B.W. Char, K.O. Geddes, G.H. Gonnet, B.L. Leong, M.B. Monagan, and S.W. Watt. *Maple V Library Reference Manual*. Springer-Verlag, New York, 1991.

[5] J. Douglas Child et al. *Calculus Laboratories for Brooks/Cole Software Tools*. Brooks/Cole Publishing, Pacific Grove, California, 1992.

[6] A. Fattahi. *Maple V Calculus Labs*. Brooks/Cole Publishing, Pacific Grove, California, 1992.

[7] K.O. Geddes, S.R. Czapor, and G. Labahn. *Algorithms for Computer Algebra*. Kluwer Academic Publishers, New York, 1992.

[8] D. Harper, C. Wooff, and D. Hodgkinson. *A Guide to Computer Algebra Systems*. John Wiley & Sons, Chichester, England, 1991.

[9] Andre Heck. *Introduction to Maple*. Springer-Verlag, New York, 1992.

[10] W. Ellis Jr., E. Johnson, E. Lodi, and D. Schwalbe. *Maple V Flight Manual*. Brooks/Cole Publishing, Pacific Grove, California, 1992.

[11] Wade Ellis Jr. and Ed Lodi. *Maple for the Calculus Student: A Tutorial*. Brooks/Cole Publishing, Pacific Grove, California, 1989.

B.2 Some research articles on Maple and its usage

[1] J. Cizek, J. Paldus, U. W. Ramgulam, and F. Vinette. Two-point Pade approximants in electrochemical kinetic currents. *Progress in Surface Science*, 25(1–4):17–39, 1987.

[2] R. M. Corless and D. J. Jeffrey. Solution of a hydrodynamic lubrication problem with Maple. *Journal of Symbolic Computation*, 9(4):503–513, 1990.

[3] R. Cushman and J. A. Sanders. The constrained normal form algorithm. *Celestial Mechanics*, 45:181–187, 1989.

[4] Garbey, Kaper, Leaf, and Matkowsky. Using Maple for the analysis of bifurcation phenomena in condensed-phase surface combustion. *Journal of Symbolic Computation*, 12(1):89–114, 1991.

[5] J. Grotendorst. Approximating functions by means of symbolic computation and a general extrapolation method. *Computer Physics Communications*, 59(2):289–301, 1990.

[6] J. D. Hobby. Numerically stable implicitization of cubic curves. *ACM Transactions on Graphics*, 10(3):255–296, July 1991.

[7] M. K. Kwong. Uniqueness results for Emden-Fowler boundary value problems. *Nonlinear Analysis, Theory, Methods and Applications*, 16:435–454, 1991.

[8] F. W. Letniowski and R. G. McLenaghan. An improved algorithm for quartic classification and Petrov classification. *General Relativity and Gravitation*, 20(5):463–483, 1988.

[9] P. K. H. Ma and W. H. Hui. Similarity solutions of the two-dimensional unsteady boundary layer equations. *Journal of Fluid Mechanics*, 216:537–559, 1990.

[10] M. B. Monagan and A. J. Granville. The first case of Fermat's last theorem is true for all prime exponents up to 714,591,416,091,389. *Transactions of the American Mathematical Society*, 306(1):329–359, 1988.

[11] R. A. Moore, T. C. Scott, and M. B. Monagan. Resolution of many particle electrodynamics by symbolic manipulation. *Computer Physics Communications*, 52(2):261–281, 1989.

[12] Barry Simon. Four computer mathematical environments. *Notices of the American Mathematical Society*, 37(7):861–868, 1990.

[13] H. Wilf and D. Zeilberger. Towards computerized proofs of identities. To be published in *Bulletin of the American Mathematical Society*.

Index

SPRINGER FOR MAPLE V